DATE DUE

OC 28 '96			
DE 5 '96			
4 '97			
DEC			
DEC 19 '97			
AP 1 '98			
AP 23 '98			
FE			
MR 14 00			
AP			
AP			
MY 29 '03			
MY 10 04			

DOUBLE-EDGED SWORD

DOUBLE-EDGED SWORD

SWORD

*The Promises and Risks of the
Genetic Revolution*

KARL A. DRLICA

§ Helix Books

ADDISON-WESLEY PUBLISHING COMPANY

Reading, Massachusetts • Menlo Park, California • New York
Don Mills, Ontario • Wokingham, England • Amsterdam • Bonn
Sydney • Singapore • Tokyo • Madrid • San Juan
Paris • Seoul • Milan • Mexico City • Taipei

Many of the designations used by manufacturers and sellers to distinguish their products are claimed as trademarks. Where those designations appear in this book and Addison-Wesley was aware of a trademark claim, the designations have been printed in initial capital letters.

Library of Congress Cataloging-in-Publication Data

Drlica, Karl.
    Double-edged sword : the promises and risks of the genetic
  revolution / Karl A. Drlica.
      p. cm.
  Includes bibliographical references and index.
  ISBN 0-201-40838-4
    1. Medical genetics—Social aspects. 2. Genetic engineering—
Social aspects. I. Title.
  RB155.D76 1994
  616'.042—dc20   94-14033
                                        CIP

Jacket design by Mike Fender
Text design by Diane Levy
Set in 10.5-point Bembo by DEKR Corporation, Woburn, Massachusetts

1 2 3 4 5 6 7 8 9 10- MA -97969594
First printing, September 1994

To Ilene, Alex, and Marc

Manuscript editor and style consultant: Deborah Everett

C O N T E N T S

Introduction **1**

CHAPTER ONE

Doctor, Is My Baby Okay? **7**

New life-or-death choices face prospective parents

CHAPTER TWO

Of Physicists and Phages **19**

Interactions of molecules explain life

CHAPTER THREE

The Golden Age **40**

DNA molecules carry the instructions for life

CHAPTER FOUR

The Enemy Within **56**

Genetic histories predict future health

CHAPTER FIVE

Fixing Bad Genes **76**

Molecular surgery attacks fatal genetic diseases

CHAPTER SIX

Early Detection of Infection **87**

Diagnosis of fatal infectious disease changes interpersonal interactions

Contents

CHAPTER SEVEN

Who Gave It to Me? *96*

Genetic methods make people accountable for spreading infection

CHAPTER EIGHT

Beyond Reasonable Doubt *108*

DNA fingerprinting enters the courtroom

CHAPTER NINE

Body Enhancement? *126*

Availability of potent biological molecules stimulates personal experimentation

CHAPTER TEN

Toward Universal Screening *134*

Genetic testing will guide future health decisions

CHAPTER ELEVEN

Society and the Revolution *143*

Ethnic diversity and big money raise social problems

Appendix I: Additional Reading 152

Appendix II: More About DNA and RNA 157

Appendix III: Family Analysis 176

Appendix IV: The Law and Genetic Discrimination 187

Appendix V: Where to Get Help 189

Glossary 219

Sources for Quotations 235

Index 237

A C K N O W L E D G M E N T S

I am indebted to many people for encouragement, editing skills, and assistance in collecting the genetic information described. They include Ardis Anderson, John Balbalis, Craig Benham, Arlene Blecher, Richard Burger, Shirley Chapin, Joanna Chin, Beverly Domingue, Diane Dresher, K.F. Drlica, Beverly Fogle, Simon Ford, Robert Gellibolian, Marila Gennaro, Paul Gottlieb, Paul Hagerman, Carol Hopkins, Holly King, Gary Kravitz, John Kornblum, Janet Kramer, Barry Kreiswirth, Ruth Levitz, Muhammad Malik, Michael Meyerowitz, Leonard Mindich, Ellen Murphy, Kathryn Neuhoff, Abraham Pinter, Steven Projan, Gianni Riotta, Barbara Rosenberg, Lance Rook, Bob Rosenbaum, Lise Rubenstein, Anne Schwartz, George Sideras, Richard Sinden, Diane Stec, Max Varon, Ilene Wagner, Jian-Ying Wang, Richard Western, and many others.

---

     †

Introduction

DNA, RNA, enzymes, ribozymes. . . . For the past ten years the popular press has peppered its news stories with molecular terminology. Until recently, most of us probably ignored the jargon, since it concerned abstract ideas having only obscure meaning. But earlier theory is now reality; genetics headlines now have relevance for the average person:

DENTIST GIVES PATIENTS AIDS

MILITARY STORES BLOOD FOR DNA IDENTIFICATION

CONVICTED RAPIST CLEARED BY DNA
FINGERPRINTING

    The genetic revolution has penetrated mainstream America.

    The revolution began about forty years ago when James Watson and Francis Crick proposed a chemical structure for the material that stores hereditary information in our bodies. Their elegant model provided a framework for understanding many complex features of organisms. During the next twenty-five years, scientists learned how to change genetic information and how to move it from one kind of organism to another. Now we are engineering bacteria for medicines, plants and animals for food, and parts of ourselves for disease cures. We've even achieved the ability to change the nature of mankind.

    Each of us, scientist and nonscientist alike, will soon face genetic problems of one kind or another. Unfortunately, few of us understand genetics well enough to do more than simply agree with our doctors when forced to take decisive action. This is because molecular genetics is a complex science normally discussed in an intimidating language of acronyms and words ending in "-ase." The concepts seem buried under mountains of jargon. *Double-Edged Sword* attempts to uncover those concepts and make them accessible to everyone.

Some readers are already struggling with hereditary disease. For them, the rapid progress of genetics holds out new hope, particularly at the level of prenatal diagnosis. But cures are still far away, and funds for the necessary basic research are scarce. Consequently, parents of afflicted children need as much information as possible to lobby effectively, to understand their options, and perhaps even to think about home remedies. (A home remedy evolved into the 1992 Hollywood film *Lorenzo's Oil,* a story about parents who developed a dietary supplement that seemed to slow a progressive, lethal genetic disease in their son. Although one case does not constitute a carefully controlled study and the general usefulness of the diet is controversial, motivated nonscientists can learn enough biochemistry to influence the medical community.)

Most of us, however, have not yet felt the genetic revolution in a personal way. When we do, we will find that the advances in molecular genetics are mixed blessings. The pace of discovery has been so rapid that we've had little time to consider the social consequences or to enact protective legislation. For us, the revolution has a duality: it carries risks along with the new health management opportunities. Our task is to minimize the risks while maximizing the benefits.

The positives and negatives come from several breakthroughs. The most obvious is our rapidly evolving ability to discover predisposition to health problems. Before long, the medical profession will have a wide variety of genetic tests for identifying individuals especially susceptible to particular ailments, including specific environmental dangers and cancer. If you happen to be in a susceptible group, you'll be advised to arrange your life-style to avoid the particular danger. For example, some people get skin cancer quickly when exposed to sunlight. They can reduce their danger by shading themselves and by staying indoors much of the time. But with health advice comes a new problem: employers and insurance companies may shun those of us known to be predisposed to disease.

A second development concerns very sensitive ways to track hereditary molecules. Criminals, kidnapped children, smuggled animals, and infectious microbes have been identified by molecular analyses. Genetic tracking can even follow hereditary diseases as they pass from one generation to the next. This enables us to single out persons harboring genetic disease and assist them with family planning. In such cases, afflicted fetuses are identified early in pregnancy when low-risk abortion can still be performed. In principle, selective abortion could eradicate a disease. However, genetic screening can also brand a family as diseased. In the past, such families were considered cursed, and the afflicted were sometimes incarcerated or put to death as witches. In modern times, it's usually careers and financial status that suffer as members of the family are deemed uninsurable or are rejected from lengthy professional training programs.

Genetic technologies may affect groups of people as well as individuals. For example, our search for defective, disease-causing genes will eventually reveal particular forms of genes that give especially desirable characteristics. Then we'll be able to identify human embryos that have favorable traits. At an early stage in development, embryo cells can be separated and each grown into a new embryo. Embryos survive freezing, so the clones can be stored until needed. Selective implantation of "gifted" embryos into prospective mothers will then produce "superbabies." When in place, embryo selection procedures are likely to be expensive, so only privileged groups or certain high-tech nations will have access. A type of genes race could ensue, with the winners acquiring enormous economic power. While this scenario is still considered science fiction by most, the necessary technology is developing rapidly.

How soon the genetic revolution will affect you or me is difficult to predict. It depends on how rapidly new discoveries are applied, on whether protective legislation is enacted, and on the particular genes we each inherited. It also depends on our individual willingness to exploit molecular genetics. However, there is little doubt that within a few years most of us will need expert help in making genetic decisions, some of which may be life-or-death judgments. Unfortunately, the "experts" will not always have the level of expertise we might like. Some may even have agendas that differ from our own. For example, experts are likely to receive a fee for each analysis, and that could encourage them to recommend unnecessary tests. Alternatively, large health management organizations may discourage costly tests that we think are necessary. Consequently, we consumers must know the right questions to ask the experts. With this book I hope to provide enough information about molecular genetics to allow the general reader to get understandable and verifiable information from genetic advisors.

Many readers already know something about genetics, having learned in school about Mendel's laws and the process of inheritance in pea plants (see "Mendel's Laws" next page). But to anticipate new opportunities and hazards, we must move that knowledge to the level of molecular structures, structures that allow us to predict an individual's future health far in advance (some diseases can be predicted with near certainty fifty years before they appear). A few technical details are introduced, but most of this book focuses on fitting concepts together, on providing a framework for evaluating new applications of genetics. The approach is to introduce specific aspects of the genetic revolution through case histories of real people confronting actual genetic problems (except when the names of these people have already been widely publicized, they are omitted or changed, both because maintaining privacy is important and because some examples are composites of several people). The histories are then followed in each chapter by commentary and practical considerations. Chapters 2 and 3 are exceptions. In these, the emphasis is on the idea that our bodies follow predictable rules, since the molecules that make up our bodies follow very precise, predictable rules. Some of these rules are described through a history of their discovery, for that makes it possible to convey a point of view as well as introduce the language and concepts of molecular genetics. For readers with specialized interests, appendices are included that provide more information about genes, some thoughts about family analysis, and a list of organizations that assist people with genetic problems.

Double-Edged Sword begins with the case that prompted me to write the book. In the early 1980s, I met a young couple who had just discovered that both husband and wife probably carried genes for a lethal genetic disease. As I watched them struggle to build a family, it became clear that other people could benefit from their experience. Indeed, their eight-year ordeal serves as an example for us all, since we each harbor a few undesirable genes.

Mendel's Laws

Gregor Mendel, an Augustinian monk living in Moravia in the 1860s, discovered that inherited features are passed from one generation to the next as if they are particles. He asserted that the particles controlling heritable traits come in pairs and that progeny (offspring) receive one member of a pair from each parent. His conclusions came from breeding different varieties of peas in his monastery garden followed by careful comparison of the characteristics of parents and offspring. When he mated tall plants with tall plants, he always got tall offspring. Likewise, when he crossed short plants with short plants, he always obtained short plants. But when he bred tall plants with short ones, he produced only tall offspring, not a mixture and not medium-sized plants. Tallness overpowered shortness; tallness was dominant to shortness. We can depict the situation symbollically by the following equation:

$$TT \text{ (tall parent)} \times tt \text{ (short parent)} = Tt \text{ (tall offspring)}$$

where T represents the dominant form of the particle (gene), and t represents a hidden, recessive form. The hidden nature of shortness became apparent when Mendel mated two of the Tt tall progeny plants. About a quarter of the offspring were short. Thus, the shortness trait must have been retained in the first mating. It was somehow covered up. After many crosses, Mendel discovered that the tall offspring of a Tt (tall) $\times$ Tt (tall) mating were not all the same: two out of three carried the concealed, recessive gene (t) for shortness. His numbers were most easily explained by offspring receiving one trait (T or t) from each parent:

$$Tt \text{ (tall parent)} \times Tt \text{ (tall parent)} = TT \text{ (tall offspring)}$$
$$+ 2 \ Tt \text{ (tall offspring)}$$
$$+ tt \text{ (short offspring)}.$$

Mendel further found that some traits are independent—the inherited particles (genes) behaved as if others did not exist. We now know that independent inheritance arises when genes are on different chromosomes,

DOUBLE-EDGED SWORD

the long, threadlike structures that carry hereditary information. Since Mendel's time, we've learned that many complicated patterns of inheritance also occur. In some cases, neither form of a gene is dominant, and matings give a "blended" result (a plant with red flowers crossed with a plant having white flowers might produce offspring with pink flowers). Other results arise from the effects of multiple genes. Nevertheless, extreme forms of many human diseases are inherited according to the simple rules derived by Mendel.

Doctor, Is My Baby Okay?

In 1983 Cathy and Bill were like many American couples—healthy, in their late twenties, and ready to have a baby. Having come from large families, both were eager to start a family of their own. Conception was easy, and Cathy prepared to put her dancing career on hold. At the time, Bill was a budding medical scientist and a bit of a worrier. As a precaution, he asked Cathy's doctor to withdraw a small amount of amnionic fluid from around their new life and send the sample to a local clinical laboratory for testing. Technicians at the lab saw only normal cells in the fluid. Everything looked fine.

In the following months, Cathy ate carefully and avoided even the occasional glass of wine. The baby squirmed and kicked as it should, but so vigorously that it rotated into the feet-down, breech position. Breech deliveries are a bit more difficult than normal ones, and Cathy's friends advised her to get the baby's head back down by standing on her own head. The gymnastics failed to turn the baby, but Cathy's doctor seemed unconcerned. Only a minor complication, he told her. Then the delivery date passed without contractions. No problem; first babies often arrive a bit late. However, sleepless nights began to exhaust Cathy. When the baby was ten days overdue, the doctor decided to proceed with a Caesarean section.

Cathy checked into the hospital and underwent the usual preparatory steps. The next morning, a nurse wheeled her into the delivery room. As a final check, her doctor placed his stethoscope on her belly and

DOUBLE-EDGED SWORD

listened for the pumping of a tiny heart. His expression clouded. He slid the instrument to another spot—and then to another and another. There was no heartbeat! The baby was dead.

Everything had gone so well right up to the last minute. Now Cathy and Bill felt helpless. Friends and family didn't provide much reassurance, nor did the doctor's curt advice to try again for another pregnancy. Cathy and Bill were afraid that the same thing would happen again. They had to know what went wrong with their baby.

The couple insisted on an autopsy for their baby, and they discovered that its digestive tract was clogged. That didn't explain much, however, since a fetus gets nutrition through its umbilical cord, completely by-passing the digestive system. The baby certainly didn't starve to death. Then a pathologist examined tissue samples taken from the baby. He couldn't be sure, but it looked as though the baby had cystic fibrosis, a serious genetic disease. In the early 1980s, the definitive diagnosis of cystic fibrosis could be made only with a living child who could be tested for salty sweat. Cathy and Bill's baby was dead, so they couldn't get a conclusive answer. But even the suggestion of genetic disease deepened their despair, since genetic disease can strike again and again.

Cathy turned all of her attention to cystic fibrosis. She and Bill would have to come to grips with this disease, with how it would affect them and any other children they might have. She became a regular at the local college library, learning as much as she could about the disease. Cystic fibrosis affects breathing and digestion to varying degrees, with most afflicted babies surviving infancy. The books told Cathy that the disease is caused by a defect in the cellular mechanism that releases chloride ions from cells. She knew that cells are the tiny units that make up our bodies, but she had to look in other books to find that chloride ions are minute particles (atoms) that each carry a negative electrical charge. As chloride ions flow out of cells, positively charged sodium ions follow to maintain electrical balance. The sodium and chloride ions then join outside the cells, where they form salt. As salt accumulates, it draws water out of the cells. This process allows some of our tissues to moisten their protective mucus, keeping it soft and pliable. In cystic fibrosis, chloride channels fail to open properly, the salt concentration outside cells does not increase, and water is not secreted. Thus deprived of moisture, the body's mucus becomes abnormally dry.

Doctor, Is My Baby Okay?

Dry mucus is a serious problem for lungs. Delicate lung tissue is normally covered by a thin fluid layer of mucus that is continually removed and replaced. Dry mucus isn't easily displaced, and so it becomes thick. That in turn provides a good place for growth of bacteria, those microscopic, single-celled organisms that sometimes cause disease. Their presence gradually breaks down the fragile lung tissue, and that is often the cause of death. Lack of water secretion also prevents the pancreas from properly releasing substances (proteins) needed for absorption of certain foods; thus, babies with cystic fibrosis often fail to thrive.

Cathy discovered that the genetic aspects of cystic fibrosis were not hopelessly against her. The disease is ultimately caused by a defective gene, one of the many units of heredity that comprise our body plans. We have two very similar, but usually nonidentical, copies of each gene. In the case of cystic fibrosis, both copies must be defective for the disease to occur. Thus, cystic fibrosis is a recessive characteristic: a person can be a carrier of the disease without being sick (this is the same idea Mendel uncovered with tall pea plants that carried a gene for shortness). Some of Cathy and Bill's babies could be healthy carriers, just as they were.

The odds of having a sick baby with the next pregnancy were one in four (Figure 1.1). The numbers emerge from the arrangement of genes in our cells. Genes are sections of long, thin molecules called DNA (molecules and DNA are defined in the next chapter). For now, we can consider DNA as a long string that is extensively folded and compacted. DNA-containing structures are called chromosomes, and at certain stages of cell growth they can be seen with a light microscope. Everyone's chromosomes come in pairs (human beings have 22 pairs plus two sex chromosomes, an X and a Y for males or two Xs for females). While each member of a pair differs slightly from the other member, the two have the same series of genes. Thus our cells contain two copies of each gene (an exception occurs with males, who have only one copy of the genes located on the X and Y chromosomes). Neither Cathy nor Bill had the symptoms of cystic fibrosis, so each of them could have only one bad gene for the chloride channel. (If either had two bad genes, that person would have the disease.) At conception, a baby receives two copies of every gene (except those on X and Y in males), one from each

DOUBLE-EDGED SWORD

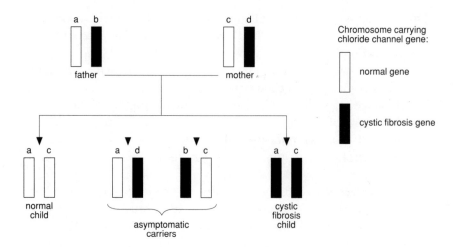

Figure 1.1. Pattern of inheritance of cystic fibrosis.

Figure adapted from M.A. McPherson and R.L. Dormer, *Molec. Aspects Med.* 12:4, 1991. Elsevier Science Ltd., Kidlington, UK.

parent. The baby's copies for the chloride channel can include either the good or the bad gene from either parent. There are only four possible combinations: a good gene from each parent, a good gene from the father and a bad one from the mother, a bad gene from the father and a good one from the mother, or a bad gene from each parent. Only the last possibility leads to disease, one case out of four. In two out of four cases, the baby becomes a carrier.

Cathy regarded the odds differently: the chance for having a healthy baby was three out of four. That would be pretty good odds in a card game. Cathy wanted a baby so badly that she would have gambled with even worse odds. As she put it, "I knew I was taking a chance, but I just ducked my head and plowed forward with another pregnancy."

The second baby was conceived about a year after the first died. In the meantime, Cathy and Bill moved temporarily to Paris so Bill could get some high-tech research training. They made friends quickly and adapted easily to French life.

Doctor, Is My Baby Okay?

From a Parisian friend, they discovered that French scientists could detect cystic fibrosis by examining fetal cells. The cells could be obtained from the fluid surrounding the baby, just as Cathy's doctor got them from the first pregnancy. This was heartening news, especially for Bill. He dreaded having a child that would be seriously ill for its entire life. An early test would allow them to avoid that by aborting a diseased fetus. Of course, the situation was still complicated—the cystic fibrosis test could only be performed at the sixteenth week of pregnancy, too late for a simple abortion if the result proved positive.

Cathy and Bill were optimistic, since the odds favored them. When the sixteenth week arrived, they drove to the doctor's office to have the amnionic sample taken. Cathy drank what seemed like gallons of water to help the doctor locate the baby by ultrasound. The baby's image showed up clearly on the TV-like screen. As the doctor stared at the screen, his face told Cathy that something was wrong. The baby's digestive tract was blocked. That wasn't a sure sign of cystic fibrosis, but in the mid-1980s it was considered a strong indicator. Without a word, the doctor picked up a felt pen and marked an X on Cathy's belly. A long needle flashed by, followed by a stabbing pain as fluid was drawn from her womb. Then came days of waiting for the test result. When the doctor called, Cathy knew immediately that the news was bad—in Paris, good news arrives by mail. "Come back to confirm it," she was told. Then came more waiting, and finally she and Bill were told, "No doubt. Cystic fibrosis. I recommend that you terminate the pregnancy."

But the baby was alive, kicking, moving around. How could you just put out its life? Cathy's mother had preached that you take what you get, and her mother lived by that rule, having raised a severely retarded child. But Cathy's mother had few options. Now, Cathy and Bill faced an agonizing decision.

Logically, Bill preferred termination, and to blunt his feelings he refused to touch Cathy's belly and feel the child kick. Cathy was torn by the decision to be made, especially since she had learned that some cystic fibrosis babies could lead short, but nearly normal, lives.

Then Bill had an asthma attack, the worst in years. He was rushed to the hospital, barely able to breathe. As Cathy watched him struggle for breath, she realized that she couldn't stand to see a child gasp that way

DOUBLE-EDGED SWORD

all day, every day. Her direction became clear: the pregnancy had to be terminated.

At week eighteen, Cathy entered a Paris maternity hospital. As she settled into her room, a calmness came over her. She had a job to do, one that she hated but one that had to be done. The French doctors worried about Cathy's scar from the earlier Caesarean section, since it was a weak spot where the uterus might tear during delivery. Induce labor slowly, they thought. And of course no anesthesia—if the uterus did tear, they'd only know if Cathy felt it and screamed in agony. A week passed before the contractions finally came. Violent contractions, six continuous hours of them. Then it was over. No immediate grief, just numbness.

Unfortunately, Cathy and Bill had no support group in Paris that could empathize with their loss. Both focused on work, with weekend touring of France serving as a diversion. The pain of loss became anger as they watched other couples enjoying their small children. It just wasn't fair that childbearing was so easy for everyone else. Even their new friends and weekend traveling companions became pregnant without complications. Cathy and Bill felt abandoned—something was wrong with a science that allowed the loss of two babies in a row.

When they returned to the United States in 1986, Cathy and Bill felt stuck. Bill would not knowingly bring a baby with cystic fibrosis into the world, and Cathy would never repeat late-term induced labor. Then Bill discovered that a new fetal test was available. The verdict could be in by the end of the first trimester, in time for a simple abortion. Cathy was not exactly overjoyed. She knew that genetic lightning could strike the same spot twice, and it was her body that would feel the life torn from it. But the alternative was no baby at all, since Bill rejected adoption. Thus a third pregnancy began. Everything seemed normal, and the prenatal sample was sent to the lab. This time the lab stretched out the wait ten days longer than promised. Ten days of absolute hell. Then the news—this baby was healthy; it had passed the test!

But could they trust the lab? Could the technicians make an error? Of course they could. Daily trips to the doctor. Please doctor, take another sonogram. The baby's not moving! Is she still alive? The first baby had slowed its movements, too. Cathy and Bill begged the doctor to stop waiting. Go in and get her out alive!

Doctor, Is My Baby Okay?

Same hospital, same staff, same delivery room. The doctor got set for another Caesarean. Bill, banished from the operating room, peered through a little window in the door. He could see only Cathy's face and the doctor's back. The doctor bent over. His elbows arched out as he reached for the baby. Then he straightened up, turned around, and held out a healthy little girl. The tests had been right after all. "I was even grateful that our baby woke up and cried every night," Cathy said later. "For eighteen months that's how I knew she was still alive."

The story does not end with one baby. Cathy and Bill wanted another child. Two undiagnosed miscarriages followed, with fetuses lost before the diagnostic test could be performed. In early 1991, the couple had a stable pregnancy. By this time, an even newer test for cystic fibrosis was available, one that could be performed very early in pregnancy. First, samples of blood were drawn from Cathy and Bill. DNA, the submicroscopic carrier of hereditary information, was removed from their blood cells and analyzed. Cathy's test result was very clear. Her DNA lacked a small bit of information in the region needed for her cells to construct the chloride channel. As pointed out previously, our cells have two similar copies of the information in DNA; only one of Cathy's DNA copies could have been defective because she didn't have cystic fibrosis. The alteration Cathy had, called ΔF_{508}, is common, accounting for the defect in about 70 percent of the carriers of cystic fibrosis (the other cases are spread among a number of different defects, all probably in the same region of DNA). The problem with Bill's DNA was not picked up by the tests. It must be there, though, since their second baby definitely had cystic fibrosis, and that required a defective gene from him as well as from Cathy. DNA was then extracted from the fetal cells. They did not have Cathy's DNA abnormality. Thus the baby must have gotten a good DNA copy from her, and that's all he needed to be free of cystic fibrosis. Six months later, Cathy and Bill had a healthy little boy: persistence and genetic testing had given them a family.

———+··———

Genetic disorders occur in roughly one of every 500 live births. Although this incidence may seem high, it is much lower for any particular disease. Cystic fibrosis is the most common Caucasian disorder—one in every twenty to twenty-five white Americans is a carrier. As in Cathy and Bill's case history, it's often the birth of afflicted children that identifies people as carriers of recessive diseases such as this. Now that DNA tests are available, screening programs can identify afflicted fetuses, and through abortion they can reduce the incidence of genetic disease.

Genetic detection programs have been very successful. A striking example can be seen among the Ashkenazic Jews of North America, a group having an abnormally high incidence of the neurological disorder called Tay-Sachs disease (1 in 3,600 births as compared with 1 in 400,000 for other peoples of the world). In this disease, newborns appear normal, but within six months they begin to lose control of their heads. Convulsions follow, the head grows abnormally large, and the child goes blind. Such children typically die before age four. To avoid arranged marriages between carriers of the disease, some religious communities have instituted premarital screening for Tay-Sachs. In less traditional circles, screening occurs after marriage; fetuses testing positive for the disease are aborted. Since 1970, the incidence of babies born with Tay-Sachs disease has dropped by 90 percent. Of course, this combination of screening and abortion doesn't rid the population of the bad gene, since babies who are carriers survive and pass the gene to their children.

We can expect the number of screening programs to increase dramatically as more disease-causing genes are identified. Some of these new genetic choices will cause problems, particularly when they have an economic impact on insurance companies. One problem case has already surfaced: a large health management program agreed to reimburse a couple for the cost of a cystic fibrosis test, but the company also insisted that if a fetus testing positive were allowed to live, the company would not cover the child's health care costs. Pressure from the medical community eventually rolled back this policy, but not before it had caused the prospective parents considerable anguish. A case of this type raises ethical questions concerning what insurance companies should cover. The answers will become increasingly important as fetuses are given batteries of tests that reveal predisposition to particular diseases, not just the certainty of disease.

The companies developing gene tests also have a big stake in screening. For a disease such as cystic fibrosis, the market can be very large, especially if company salespeople convince physicians to routinely screen all patients. Pressure from the scientific community blocked an early attempt at massive cystic fibrosis screening: it was deemed irresponsible to let a patient feel safe from disease when the test missed 30 percent of the carriers. Technology is now available to pick up all changes in the cystic fibrosis gene (more than 300 deviations from normal have been found), and that raises another problem: many of the deviations are not associated with a clear manifestation of disease. Even DNA alterations known to be serious occasionally fail to elicit disease, as if another gene also participates in cystic fibrosis. What originally seemed to be a simple problem is now quite complex.

Complex issues are also associated with newborn testing. For babies already born, the main consideration is that the infants benefit from screening. Such is certainly the case for tests that identify those with phenylketonuria, since special diets prevent the mental retardation associated with this disease. However, newborn screening programs are often flawed. For example, it does a baby no good to be identified as a carrier of a recessive disease such as sickle cell anemia. Indeed, such information can be harmful if the child becomes branded as a disease carrier. This general problem with newborn testing is widely recognized, but screening programs in many states continue to reveal carrier status of babies. Other tests, such as that for maple syrup urine disease, are applied even though of questionable value: frequently the test result for this rare, fatal disease is not available until after the baby is seriously ill. Problems with newborn screening are likely to increase as new DNA tests become available. As a result, the National Academy of Sciences recommended in 1994 that health departments evaluate the risks and benefits of screening much more carefully. Particularly important are the blood samples taken from babies, since these are stored for future use without informed consent from the tissue donors.

———+··

The question most people ask about genetic testing is whether it is worth the trouble and expense. You can make a good guess by thoroughly examining your family medical history. Senior family members usually know who died of what; quizzing them can turn up inherited diseases. Even a little information can be valuable. For example, a colleague of mine, after hearing his sister mention that she was starting a family, reminded her that one of their cousins had suffered from Tay-Sachs disease. This comment led to testing of the sister and her husband, with the discovery that both were carriers of the disorder. Prenatal testing of the first baby showed that it would be affected, and that fetus was aborted. The second baby was healthy. The third miscarried; it tested positive for Tay-Sachs disease. The fourth was healthy.

You can begin your own analysis by sketching a family tree and noting diseases next to each family member. Look for anything unusual. A few of the more obvious disorders are mental retardation or illness, infertility, miscarriages, and early death. Even infectious diseases, if serious, are worth recording, since predisposition can be inherited. The appearance of patterns will give you more specific questions for your relatives and perhaps a reason to consult a genetic counselor. As a guide, the general categories of genetic disease are discussed in Appendix III.

Your collection of genetic information should be recorded and stored with your personal records. Seemingly unimportant details may become very valuable genetic history for your children, grandchildren, and great-grandchildren. Since genetic information can potentially be used against you and members of your family (see Chapter 4), your findings should be kept strictly confidential. Obviously, the less you rely on outside help, the more you control the information.

Professional help can be obtained from agencies listed in Appendix V. These organizations often provide reading lists concerning diagnosis and treatment of particular diseases. It may be wise to consult a good genetic counselor (your family physician may not be on top of this fast-moving field). Professional counselors, well-versed in the DNA tests and their emotional implications, can be found through regional genetic services listed in Appendix V. Although genetic counselors are trained to explain complex genetic concepts to the general public, you can greatly increase the value of the sessions by being familiar with the language of molecular genetics. Chapters 2 and 3, Appendix II, and the glossary of this book will help with DNA terminology. Since the jargon can get pretty thick, you shouldn't count on remembering everything. Even when written reports are provided, it may be useful to tape record the session. In the end, you should expect to understand the test and the disease well enough to weigh the positive and negative aspects of knowing the future (some negative aspects are discussed in Chapter 4).

Even if no obvious history of disease emerges, some prenatal screening is still advised if previous miscarriages have occurred, if maternal age exceeds thirty-five years (chromosomal abnormalities in the mother increase with age), or if either parent or previous offspring have chromosomal abnormalities. But screening shouldn't be considered routine for all pregnancies—there is a small risk of miscarriage or fetal damage associated with obtaining the fetal cell sample. In general, the risk increases the earlier the cells are taken during pregnancy. Since this risk varies from doctor to doctor, knowing a particular doctor's miscarriage rate, or at least knowing how many times the physician has performed the procedure, will be helpful in selection.

Termination of pregnancy is not the only solution for genetic disease. In some cases, a special diet for the baby will correct the condition. For others, frequent hospitalization can be expected. Then the extent of insurance coverage may become a factor in the abortion decision. To help with the choice, it is often useful to contact parents who already have "special" children. They can provide insights into the day-to-day problems and joys. Appropriate individuals can be located through the list of organizations in Appendix V.

Finally, it should be pointed out that Cathy and Bill were well versed in molecular genetics, and this allowed them to maintain some objectivity. For example, they knew that their genetic problem was nobody's fault and that nothing Cathy or Bill did during her pregnancy increased the fetus's risk of disease. They also knew, at least intellectually, that carrying a gene for a recessive disorder doesn't make a person defective. Indeed, many of the genes responsible for recessive diseases may actually provide a health advantage under certain conditions (for example, sickle cell trait, which is found in more than 2.5 million Americans of African and Mediterranean descent, helps protect against malaria).

The more I learned about Cathy and Bill, the clearer it became that an understanding of genetics is important for exploiting new medical developments. They asked the right questions, and they knew what to push for. To give you a grasp of the fundamentals of genetics, the next two chapters sketch the origin of the genetic revolution. Chapter 2 focuses on Max Delbrück's study of bacteria and viruses, while Chapter 3 outlines the activities of DNA uncovered in the golden age of molecular biology. These chapters introduce many of the concepts developed in later chapters, and at the same time they provide some of the clearest examples of basic research driving medical science. The early molecular biologists gave us a strong sense that life can be explained in terms of tiny, lifeless particles called molecules. From that perception emerged the guiding assumption of the genetic revolution: the workings of our bodies are often explainable in terms of the same molecule types that run the lives of all other organisms on earth.

Practical Considerations

- Older expectant mothers are advised to have fetal cells checked for chromosomal damage.
- If pregnant, begin assessing your baby's risk of genetic disease by making a family tree that lists early deaths, diseases, miscarriages, and other problems (see Appendix III). Keep this information private.
- Although DNA analysis may show that you carry a particular variation of a gene, the ramification for your child may be much easier to predict if your family's medical history is available.
- If pregnant, learn what mandatory tests will be run on your baby. Regulations vary widely from state to state, with some states even testing for untreatable diseases and revealing carrier status (either could stigmatize your baby). If you have a choice about which tests will be performed, learn which ones are really important and why. Try to ensure confidentiality of test results.
- A blood sample from your baby may be saved for future DNA analysis. Be certain that the genetic information obtained cannot be used against you or your baby before signing a release.
- Before elective genetic testing, consider risks such as uncovering parentage secrets (cases in which the identity of a biological parent has been hidden from a spouse or child).
- Before elective genetic testing, consider insurance ramifications. Inclusion of family information in an insurance database may lead to exclusion of future benefits.

CHAPTER   TWO

---------------------------- † ----------------------------

Of Physicists and Phages

In the 1930s, Max Delbrück, a young German physicist, began to wonder how heredity works. Simple formulas for predicting inheritance had emerged from breeding experiments by geneticists, and individual genes seemed to be arranged one after another in a row. But at the time, genes themselves were only abstract, logical constructs, much like the *T*'s and *t*'s used to describe Mendel's experiments with peas (see Introduction). What genes are made of, how they reproduce, and how they can suddenly and heritably change (mutate) were still great mysteries.

By 1932 Delbrück had completed his Ph.D. research in quantum physics and had established solid credentials during three additional years of study in England and Denmark. While abroad, he mingled with biologists who talked incessantly about inheritance. Their questions gnawed at Delbrück after he returned home to study nuclear fission. He made several important contributions to physics, but he couldn't keep away from biologists. A Russian geneticist gradually captured Delbrück's spare time with discussions of heredity, and soon the pair was joined by another German physicist. Together they struggled to define the essence of life. They went on study binges, sometimes skipping work for days. Finally the little group reduced life to an "evolutionary accumulation of experiences." From that perspective, understanding life meant determining how living matter changes to perpetuate the experiences—that is, how genes change and then duplicate to pass the change from one generation to the next. For Delbrück, all other scientific problems lost importance.

19

By 1937 the Nazis were making life difficult for many scientists. Delbrück could not mask his disdain for the Nazis, and they knew he'd never be a good party member. When the Rockefeller Foundation suggested that the tall, precisely speaking German pack his bags and move to America, Delbrück eagerly accepted their offer of a biology fellowship. At the time, his best hope for understanding how genes change was to study fruit fly mutations at Caltech. So off he set for Pasadena.

Soon after Delbrück arrived in California, he discovered that the amount of fly lore to be learned was massive. He doubted that he could ever absorb enough fly genetics to do an intelligent experiment. His interactions with fly geneticists didn't help. He was regarded as an outsider, a theoretical physicist trying to tell geneticists how to do their jobs. At his first seminar, his "Let us imagine a cell as a homogeneous sphere" drew laughter, not serious thought. Had Hitler not been so powerful, Delbrück probably would have returned to Germany.

Earlier frustrations with science had shown Delbrück the value of flexibility. His first love, astrophysics, had demanded mathematical skills that were described only in English, which at the time he couldn't read. So he moved on to the German-dominated field of quantum physics. Soon after moving to Caltech, he discovered Emory Ellis. Delbrück immediately gave up on fruit flies.

Ellis was a virologist who had been relegated to a basement lab, where he worked on an offbeat study of bacterial viruses (bacteria are microscopic, single-cell organisms; viruses are tiny, infectious particles that will gradually be described here; those that attack bacteria are called bacteriophages, or phages for short). In the 1920s, phages had generated interest as possible cures for bacterial diseases, but that hope had fizzled when no breakthroughs materialized. Now Ellis thought phages might help him understand cancer, since some viruses cause cancer in animals. In that hope, he was pretty much alone.

Delbrück visited Ellis mainly out of curiosity. He had almost given up his dream of merging theoretical physics with genetics, and he didn't expect any great insights from his trip downstairs to see Ellis. However, Delbrück had a general interest in viruses, since he considered them to be little more than genes that reproduce. He had even toyed with the idea of studying plant viruses until he discovered how inefficiently they infect their host cells. Their numbers couldn't be reliably counted, so they were of no use to him. Delbrück had hardly finished shaking Ellis's hand when Ellis showed him a trick for counting phages (phages are so small that they can't be seen and counted using ordinary microscopes). By spreading millions of bacteria on a dish filled with agar, a gelatinlike substance, Ellis could get the bacteria to grow and divide until they formed a tightly packed lawn that completely covered the agar. If one phage particle were added to the bacterial culture when it was first spread on the plate (before the lawn formed), the phage would infect one of the bacterial cells and produce offspring. That would kill the bacterial cell. The new phages, which cannot swim or breathe or do anything on their own, escape when the dead bacterial cell breaks apart. They would then attach to neighboring bacterial cells, penetrate those cells, and cause still more phages to be made as they killed the bacteria. Quickly the zone of dead bacterial cells would expand on the agar surface. After half a day, the bacteria surrounding the death zone would become too densely packed to keep growing and dividing. At that point they would no longer support phage reproduction, but a hole would remain in the lawn where the phages had killed bacteria. Such a hole, called a plaque, is easily seen with the naked eye (see Figure 2.1). To determine phage numbers, Ellis simply looked at an agar plate and counted the plaques in the bacterial lawn.

The simplicity and elegance of Ellis's work captivated Delbrück. Ellis had an experimental system that was beyond Delbrück's wildest dreams. Results could be collected in a day instead of the weeks or months needed for fly experiments, and phage data could be analyzed mathematically. Together he and Ellis could conduct simple biological studies comparable to physics experiments on atoms!

Delbrück's first phage work was a careful, quantitative study of what happens when phages infect bacteria. He and Ellis infected hundreds of millions of bacteria simultaneously with phages by simply mixing them together. Then they removed samples at various times to count the number of phages present as infection proceeded (samples of infected bacterial culture were added to new bacteria, placed on agar plates, and incubated overnight to see how many plaques would appear). The number of phages present in the infected culture did not change until after about thirty minutes, and then it quickly jumped almost a hun-

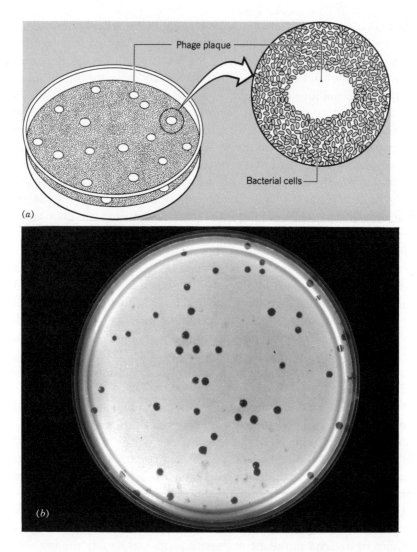

Figure 2.1. Bacteriophage plaques.

An agar plate is covered by a lawn of bacteria. The holes in the lawn, called plaques, are regions where phages have killed the bacteria. (a) Schematic diagram. (b) Photograph of agar plate. (Photograph courtesy of Robert Rothman. Figure taken from *Understanding DNA and Gene Cloning* by Karl Drlica, John Wiley & Sons, New York).

dredfold. It was as if new phages were being made inside the bacteria during the first thirty minutes, and then the phages were released in a burst, hundreds per cell, billions per flask. If Delbrück could figure out what happened in that first half hour, he would understand the essence of life, for that was when the phage genes reproduced.

Delbrück's fellowship ended in 1939, and it looked as if his phage work would halt (Ellis had already returned to real cancer studies). Since Delbrück's academic future in Germany was bleak (he had twice failed to show sufficient "political maturity" at a Nazi indoctrination camp), he was reluctant to return. He survived for several months by borrowing money from friends, and then the Rockefeller Foundation rescued him a second time. The Foundation persuaded Vanderbilt University to give him a physics instructorship (the Foundation was actively grabbing European scientists for American universities, and Delbrück was considered a very bright theoretical physicist). The Foundation agreed to pay much of his salary, and that allowed him to continue his phage work.

At about the same time, Salvadore Luria, a young Italian war refugee in New York City, also began to experiment with phages. Luria admired an old paper that Delbrück's Berlin study group had published, and he had tried unsuccessfully to join Delbrück at Caltech. Luria wanted to determine the chemical structure of phage genes, a desire that complemented Delbrück's own effort to understand how phage genes reproduce. The two first met in 1940. Although Luria was a microbiologist, he had enough training in physics to talk Delbrück's language. The two worked well together and soon produced an elegant study showing that bacteria, just like all other organisms, have genes that can spontaneously mutate. This conclusion, which was contrary to the prevailing view of bacteria, started the fields of bacterial genetics and molecular biology.

By 1943 Delbrück and Luria had attracted the attention of another microbiologist, Alfred Hershey. The three formed a loose cooperative, the Phage Group, that shared data and ideas. Collectively, they expected to learn why children look like their parents. To avoid dilution of effort, Delbrück, either by edict or negotiated treaty, persuaded the others to focus on a few of the many phages that infect the colon bacillus (*Escherichia coli,* or *E. coli* for short). This decision helped make *E. coli* and its viruses the best-understood life forms on earth.

Before he left Berlin for California, Delbrück had published a theoretical consideration of genes and mutations. The work gathered dust for almost a decade, although he had sent copies of the paper to the many prominent physicists he knew. Eventually the work caught the attention of Erwin Schroedinger, one of the best-known theoretical physicists of the time. During the war years, Schroedinger had set up shop in Ireland, where, as a condition of employment, he presented an annual series of public lectures. For one set he expanded on some of Delbrück's early ideas, and in 1944 he published the lectures as a short book called *What Is Life?* The book had almost no impact on biologists, but Schroedinger's status gave Delbrück a strong constituency among physicists, who considered the book important reading. Perhaps this attention arose because many famous physicists were looking for a new kind of work after the war and the bomb, one they could control. Or perhaps the attention arose because Delbrück and Schroedinger convinced other physicists that the search for the gene would reveal new physical principles. They thought they needed a new idea to explain how something as small as a gene could, as a single unit, be so stable and predictable while all other small particles, such as atoms, were predictable only when averages of very large populations were considered. Regardless of the reasons, Schroedinger's book drew physicists to Delbrück and his phages.

Delbrück also recruited. He had originally thought that within a year he would understand how phages replicate, but after eight years of struggle, the riddle of life was still unsolved. "Well, I made a slight mistake," he said during a prestigious Harvey lecture in New York City. "I could not do it in a few months. Perhaps it will take a few decades, and perhaps it will take the help of a few dozen other people. But listen to what I have found, perhaps you will be interested to join me."

In 1945 he began bringing new blood into the phage business with a summer course at the Cold Spring Harbor Laboratory on Long Island. This picturesque setting of beaches and wooded trails was perfect for the combination of walking, talking, and thinking that often leads scientists to new concepts. During the nine-day workshop, Delbrück gave physicists clear definitions of the phage system and a simple set of experimental techniques. By the end of his boot camp, the recruits were ready to embark on their own; no further preparation was required. But more important was the spirit Delbrück fostered as he generated a collective effort to uncover the nature of genes. The net effect was that 30 out of the 130 students of the first ten courses became internationally recognized in their new field.

Delbrück's own commitment was complete, and he tolerated no dilettantes (he made all participants in the course pass a mathematics entrance examination). But he also had a childish love of practical joking, and to everyone he was just Max: caustic critic and chief cheerleader. For twenty-six years the workshop attracted new talent to phage and bacterial genetics.

The phage physicists were comfortable experimenting with things too small to be seen. Indeed, they had spent most of their careers trying to understand the invisible forces that operate within atoms (atoms are the tiny particles that make up all substances in the universe; see "Atoms and Molecules," below). By perturbing their experimental subjects, physicists had been able to describe atoms by simple and precise rules. For example, they had learned that every atom is composed of a dense, positively charged central core, a nucleus, surrounded by an organized system of negatively charged electrons. Amazingly, the hundred or so atom types that make up the entire universe differ mainly in the number of positively charged protons each contains in its nucleus. Thus elements, such as oxygen, hydrogen, silver, and gold, are each composed of atoms having a unique and characteristic number of protons (hydrogen has one proton, oxygen eight, gold seventy-nine, etc.). The physicists reasoned that if all matter, including living things, is made of atoms and if atoms behave according to precise rules, then life must also behave according to a set of rules. Moreover, there must be a continuum within the spectrum of life—the simple rules that make progeny phage like their parents must also make human babies like their parents.

Atoms and Molecules

If you use a magnifying glass to look closely at a picture in a book, you will see that it is composed of tiny dots. Together the dots make a solid image. Likewise, every substance in the universe is made of tiny particles called atoms. A substance composed of a single type of atom is called an element. There are roughly a hundred different elements. Substances can also be composed of tightly bound combinations of atoms called molecules. In general, molecules are very small particles, although some can be quite long. The tiny particles, atoms and molecules, are moving constantly. In a solid, the particles are held together tightly; thus, individual particle movement is small. This makes solids hard and rigid. In liquids, the forces holding the particles are weaker, allowing some slippage and flow. The interactions are still weaker in a gas. There the particles dash around wildly, filling any space. Our bodies are composed of molecules that are constantly moving, usually over such short distances that we don't notice the movement.

The starting point was the idea that atoms join with other atoms to form specific combinations called molecules. Atoms in molecules are bonded together very tightly, so molecules are generally stable, discrete structures. The general concepts can be seen by considering a child's construction toy composed of interlocking plastic parts. The parts, which are analogous to atoms and which may come in a hundred different sizes and shapes, can attach to other pieces in only a few different ways, the equivalent of the specific rules for joining atoms. By putting many pieces together, a great variety of complex structures can be built, and these structures remain intact until the pieces are pulled apart. When the structures are disassembled, the parts can be reused to build new structures. So it is with atoms and molecules.

From Max's molecular point of view, a virus such as a phage is simply a specific set of molecules bound together. The power of a virus to reproduce depends on its molecules, not on magic. At the same time, a phage depends on its bacterial host for life processes: outside the bacterial cell, the conglomeration of molecules called a phage is as dead as salt (as far back as 1935 Wendell Stanley had shown that he could make crystals out of an infectious plant virus, just as if it were salt). Bacteria, on the other hand, are self-sufficient cells in which a specific collection of molecules is separated from the surrounding environment by a membrane, a covering that is itself comprised of specialized molecules. These living bags of molecules reproduce as discrete units that grow and then split in half. Max reasoned that since phages cannot reproduce on their own, they must carry into the bacterium a molecule, the phage genetic material, that takes command of the bacterial machinery and causes it to reproduce the phage molecules.

Max and the Phage Group approached the problem of phage reproduction in a variety of ways. One involved mutations. In the 1940s mutations were known to be sudden, inherited gene alterations that permanently changed the traits of organisms. Thus mutations provided a way to perturb genes; by studying the deviant behavior of a phage mutant, one could guess at the normal function of the altered gene. Obtaining mutants was not difficult. Some formed abnormally large or small plaques, while others grew only in certain bacterial strains. Sometimes a bacterium would suddenly and permanently become resistant to phage attack (such bacterial mutants could be seen growing in the middle of plaques). Then phage mutants would arise that could infect these "resistant" bacteria. The phage mutants were studied in a variety of ways to eventually relate specific genes with specific phage structures and activities.

As Max examined phage mutants, he found that when he infected the same bacterial cell with two different mutant phages, among the offspring were phages that lacked either mutation. Somehow the phage genes formed new combinations. The process, which is called recombination, involves breakage and rejoining of the genetic molecules so two phages can swap parts (Figure 2.2). Max's discovery left little doubt that phages possess the universal principle of heredity, since genes in higher organisms had also exhibited the ability to recombine (recombination is important in some human genetic tests, as pointed out in Chapter 4). Moreover, Max and his colleagues could use recombination and quantitative analysis to determine the relative order of genes, since new combinations form more frequently as the distance between two genes increases. By working with billions of phages, very rare recombinants could be easily detected, and that quickly moved phage work ahead of fly genetics.

Electron microscopy was also used to study phages. In the early 1940s Max and his associate Tom Anderson saw that some phages have heads and tails (Figure 2.3). By studying bacterial cells infected with phage mutants, other workers later saw the accumulation of phage parts. Some mutant phages failed to make certain parts, and that led to the assignment of roles for many genes.

A third approach, used more in later years, involved dissecting the phages into their component parts to figure out how each part works, just as you would study the organs that make up a frog or a human.

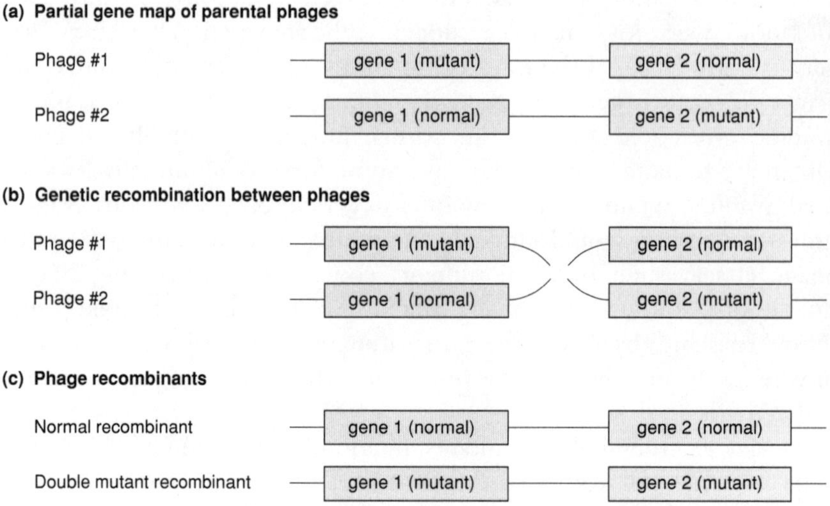

(a) Partial gene map of parental phages

| Phage #1 | gene 1 (mutant) | gene 2 (normal) |
| Phage #2 | gene 1 (normal) | gene 2 (mutant) |

(b) Genetic recombination between phages

| Phage #1 | gene 1 (mutant) | gene 2 (normal) |
| Phage #2 | gene 1 (normal) | gene 2 (mutant) |

(c) Phage recombinants

| Normal recombinant | gene 1 (normal) | gene 2 (normal) |
| Double mutant recombinant | gene 1 (mutant) | gene 2 (mutant) |

Figure 2.2. Phage recombination.

(a) Partial genetic map of parental phages. Of the two phage types used to infect the bacteria, each contains a single mutation, but in different genes. Thus neither can reproduce in the bacterium.

(b) Breakage of genetic material during infection. The genetic molecules of the parental phage types occasionally break. Although break points occur at different locations on different molecules, only one is shown.

(c) Partial genetic map of recombinant phage. The broken ends of the genetic molecules occasionally join with a different molecule and generate one that has no mutation. A phage containing this molecule can reproduce and generate a plaque.

However, in the submicroscopic world, we can't see anything directly to take it apart, so we use chemicals as our dissecting tools. Since the forces that join together the atoms of an individual molecule are often much stronger than those that hold molecules together, biologists were able to acquire chemicals for loosening the holds between molecules. Then electric currents and centrifugal force could be used to separate

Of Physicists and Phages

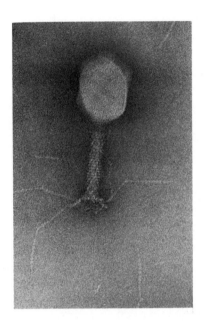
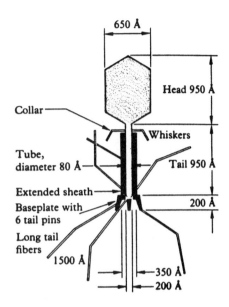

Figure 2.3.   Structure of a T-even bacteriophage particle.

The electron micrograph on the left (taken by Eduard Kellenberger) is interpreted in the sketch on the right. Numbers indicate dimensions in angstroms (an angstrom is one hundred millionth of a centimeter). The figure is reproduced from *Molecular Genetics* by G. Stent and R. Calendar, W.H. Freeman and Company, San Francisco).

and define the different molecule types. It turned out that almost all the molecules of a phage particle are very large, with the largest containing millions of atoms. By comparison, water molecules contain three atoms, oxygen gas two, and table sugar forty-five.

Analysis of these large macromolecules by biochemists revealed that they are composed of many smaller molecules bonded together. The small molecules are often similar or even identical; these repeating parts of large molecules are called subunits. The nature of the subunits made it possible to guess how the macromolecules behave and even how they are made (often the subunits are simply stuck together, one after another, to form chains). The subunit concept fit with one of Schroedinger's ideas. He had postulated that information carried by genes is arranged in code, which he illustrated with the simple example of the dots and dashes of the Morse code. The chemical subunits, bound together in chains, could serve as the dots and dashes of the genetic code.

In 1947 Caltech lured Max back as a professor of biology. His laboratory became the center of the phage world, and by the early 1950s the Phage Group had expanded to about thirty participants. Gunther Stent, one of Max's disciples, aptly described him as "a kind of Gandhi of biology who, without possessing any temporal power at all, was an ever-present and sometimes irksome spiritual force. 'What will Max think of it?' had become the central question of the molecular biology psyche . . ." and his "laboratory at Caltech . . . became the Phage Group's 'Vatican,' where many of the first generation of molecular geneticists took their orders."

By 1950 it was clear that phage genes, whatever their nature, penetrate bacterial cells, reproduce there, and direct the formation of hundreds of new phages per cell. Molecular dissection had revealed that these viruses are composed of two types of molecule: protein and DNA (both will be described later). One or both must be the genetic material. Since both protein and DNA are composed of subunits, either could have been the information carrier postulated by Schroedinger.

Bacteria also contain protein and DNA, and as first shown by Luria and Max, bacteria acquire heritable mutations. As far back as 1944, Oswald Avery and his associates had reported that DNA, when extracted from one type of bacterium and mixed with a second, gave properties of the first to the second. At first glance this would seem to prove that DNA is the molecular carrier of genes. But Avery's discovery was ahead of its time. In the 1940s DNA was erroneously thought to be a molecule of monotonous repeating subunits that was unable to store information. Without a proper framework, many scientists were skeptical that DNA was genetic material. Instead, they argued that Avery's extraction procedure might have failed to remove all the protein from the sample: a small amount of contaminating protein, rather than DNA, could have carried the genetic information from one bacterium to another. Members of the Phage Group believed in Avery's discovery, but they did not know how to use it to better understand reproduction.

Hershey, the third cofounder of the Phage Group, had noticed that an empty outer shell of a phage is left behind after the phage attacks a bacterium. It was as if a hypodermic syringe had injected something into the bacterium and then been discarded. That something must be the phage genes, since those genes had to enter the bacterial cell before progeny phage could be made. Other studies had shown that the composition of protein and DNA differ (DNA has phosphorus but no sulfur, whereas protein has sulfur but little detectable phosphorus), and that allowed Hershey to demonstrate in 1952 that phage DNA, not protein, enters bacterial cells. (When the phage particles contained radioactive phosphorus in their DNA, radioactivity could be recovered from the attacked bacteria, but none could be recovered when phages contained radioactive sulfur in their protein.) Thus DNA must carry _all_ the information for making protein and for making itself. But Hershey's phage experiment did not reveal how that could happen.

At about the same time, a group of physicists was devising methods for determining relative locations of atoms in molecules. They knew that when a beam of X-rays is directed at a preparation of molecules, some of the X-rays bounce off atoms in a way that reflects the positions of the atoms in the molecules. If film is placed behind the sample, a pattern of spots is seen on the film where the X-rays hit. The pattern is usually very complex, since there are many, many atoms in something like DNA. But when a molecule has a repeating structure, a distinct and sometimes simple pattern emerges. Then attempts are made to construct a model of the molecule to explain how the X-rays can bounce off the atoms and produce the pattern. The more closely the model predicts the X-ray behavior, the better the model describes the structure of the molecule.

One of the young scientists learning the X-ray approach was James Watson. He had started as a birdwatcher, but chance, Schroedinger's little book on life, and a great admiration for Max drew him into the Phage Group. During his student days, Watson struggled to understand how X-rays kill phages. He then bounced between a couple of Danish labs before Luria and Max persuaded him to move to England, where X-rays were being used to study the structure of molecules. The Phage Group had focused Watson's attention on finding the structure of genes, and by the time he began work in England, he was quite certain that

DOUBLE-EDGED SWORD

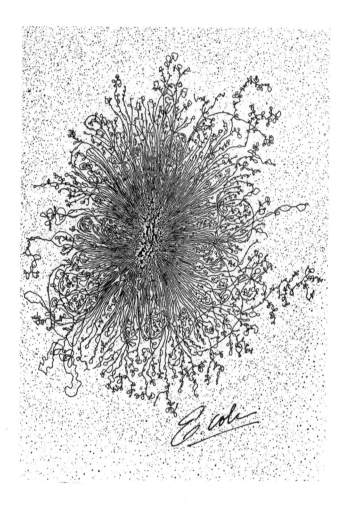

Figure 2.4. Electron micrograph of a single DNA molecule in a chromosome.

The long, threadlike material is the single, uninterrupted DNA molecule from a single bacterial cell. In this example the bacterial chromosome was removed from the cell, purified from other cellular parts, and then its DNA was spread on a surface prior to examination with an electron microscope. Prepared by Ruth Kavenoff and Brian Bowen. "Bluegenes 1" © with all rights reserved by DesignerGenes Posters, Postcards, T-shirts, etc., P.O. Box 100, Del Mar, CA 92014.

genes are made of DNA. It was obvious that he needed to define the structure of DNA.

Watson soon met Francis Crick, a physicist who had also been drawn to biology by Schroedinger's book. Crick taught Watson about X-ray analysis, and together they began to work seriously on a metal model for a short stretch of DNA. They fit together little pieces representing groups of atoms, always keeping the distances and angles between the "atoms" as accurate as possible. After a number of false starts, they hit upon a model that conformed with all the existing knowledge, including X-ray data that Rosalind Franklin and Maurice Wilkins had recently collected. The structure was elegantly simple, and when Watson described the idea to Max, he immediately recognized its significance. For him, their model of DNA equaled the discovery of the nucleus of the atom.

Watson and Crick published their description in 1953. Although it was only an idea, the model quickly transformed the study of biology, since it made obvious how a DNA molecule could reproduce, how life could reproduce. Without question, this was the biological discovery of the twentieth century.

———+·———

DNA, as currently envisioned, is an incredibly long, threadlike molecule (Figure 2.4). In bacteria, it's a thousand times longer than the cell in which it resides. (DNA from one of our cells is about as long as we are tall. If DNA from all the cells in one person were joined end to end, it would stretch to the sun and back hundreds of times.) DNA is composed of two helical, intertwined strands (Figure 2.5a), and that gives it the popular name double helix. Each strand is composed of subunits (Figure 2.5b), just as predicted by Schroedinger. Thus, we can think of DNA as two strings of beads. Each bead (subunit) is called a nucleotide. If we consider only a short stretch of DNA, on the order of hundreds of nucleotides (bacterial DNA has some eight million nucleotides; human DNA has a thousand times more), the molecule is fairly rigid. We can imagine it to be a cylinder with the two strands spiraling up the outside of the cylinder.

Each nucleotide is composed of three parts: one is a flattened ring of atoms that points toward the center of the cylinder, and the other two

DOUBLE-EDGED SWORD

form a portion of the continuous backbone of the strand (Figure 2.5b). The flattened rings are called bases, and they lie with their flat sides perpendicular to the long axis of the cylinder, much like steps of a spiral staircase. The bases stack on each other and provide a slight stiffness to the DNA molecule.

In preparing their model, Watson and Crick knew that DNA contains four types of nucleotide, each having a different arrangement of atoms

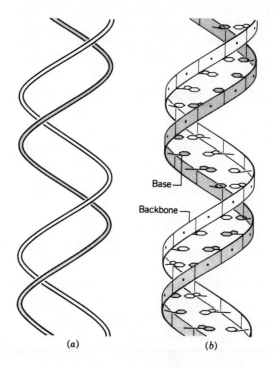

Base

Backbone

(a)   (b)

Figure 2.5.   DNA as a two-stranded molecule.

Schematic representations of DNA are shown.

(a) Two interwound strands.

(b) Double helix of two interwound strands with bases on each strand pointing toward the other. (Figure adapted from *Understanding DNA and Gene Cloning* by Karl Drlica, John Wiley & Sons, New York.)

in its base. (The bases are called adenine, guanine, cytosine, and thymine, which are abbreviated by the letters A, G, C, and T. The names of the nucleotides are derived from the names of their bases, and so the nucleotides are abbreviated by the same four letters used for the bases.) Watson and Crick also knew that DNA did not have a random assortment of nucleotides: experiments had shown that for every T in DNA, there is one A, and for every G, there is a C. But if DNA carried a vast array of information through nucleotide sequence, then the order of nucleotides must be highly varied. The A:T and G:C relationship would cut down on the number of possible sequences. Watson and Crick solved this problem by proposing that the A:T and G:C rules reflected a relationship between the two strands of the double helix. Whenever one strand contained an A, then the other would have a T opposite it. Likewise, G and C must be paired, one in each strand. Thus nucleotide sequence could vary enormously along the length of DNA. The paired bases (A:T and G:C) fit together like locks and keys, and so would the two strands. The A:T and G:C relationship also made the sequence of one strand correspond to that of the other. Consequently, if you know the nucleotide sequence of one strand, you automatically know it in the other. This idea, which is called complementary base-pairing, is probably the most important concept in molecular biology. As will be explained later, it is the basis for our current procedures for gene screening and DNA-based identification of criminals.

Complementary base-pairing explained in a general way how DNA could be copied: the two strands come apart, and then each serves as a template for the formation of a new strand (Figure 2.6). For example, wherever there is an A in the old strand, a T is placed opposite it in the new one; G in the old strand requires a C in the new one. This process, called DNA replication, results in two double-stranded DNA molecules being created from one old, double-stranded DNA. Both new molecules have exactly the same information as the old one. When cells divide, each new cell gets a complete copy of the DNA, the genetic blueprint (Figure 2.7).

The general idea of DNA replication solved the riddle of life to Max's satisfaction. Other questions, such as how information in DNA is used

DOUBLE-EDGED SWORD

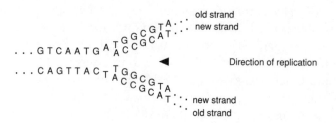

Figure 2.6. Two DNA molecules arise from one.

Complementary base pairing allows information to be copied exactly. For clarity, the strands have not been drawn as an interwound helix.

to form phage proteins, were minor details that could be left for other people to discover. Max himself set off to learn how nervous systems work.

———+·———

With the revelation of DNA structure came the official birth of molecular biology (the term had been coined almost two decades earlier, but it had not caught on). Now members of the Phage Group and others studying informational molecules realized that they were in fact molecular biologists—the secrets of life were being revealed at the level of molecules, not subatomic particles, atoms, cells, organs, or organisms. Within a few years, departments of molecular biology emerged at major research universities. Many of the faculty were Max's disciples, and they fixed his imprint on the young science. Since his style still guides some of the best work, a few more words about Max will help the general reader understand the people whose curiosity is driving the genetic revolution.

Max took his leadership role seriously. His job was to weed out bad science and encourage good. His dictum was simple: phrase precise hypotheses and then ask specific questions that can be answered. Don't collect data for its own sake! If experiments don't help answer the deep questions, then they aren't worth doing. Likewise, facts that don't help

Of Physicists and Phages

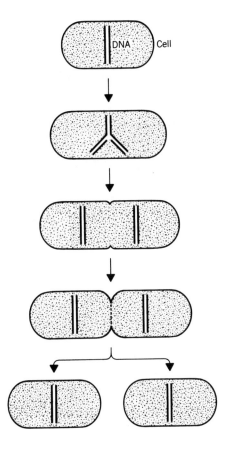

Figure 2.7. Replication and segregation of DNA.

After a DNA molecule replicates (duplicates), the two copies move apart. Cell division occurs between the copies, so each daughter cell receives the same genetic information as contained in the original cell. (Figure from *Understanding DNA and Gene Cloning* by Karl Drlica, John Wiley & Sons, New York).

DOUBLE-EDGED SWORD

you think about deep questions only clutter your mind. At times, this attitude made Max and the old guard physicists seem dogmatic. For example, Max is said to have refused to read about experiments concerning the origin of life until someone could come up with a recipe that would tell him to "Do this and do that, and in three months things will crawl in there." But his attitude did keep the effort focused.

Max was tough on his students. Neville Symonds recalls the grand master loudly interrupting him in the middle of a talk with, "Let me know when you get out of the kitchen sink [and into something interesting]." Then Max pulled out his newspaper and ignored the rest of the seminar. This stunt was so common that a whole class of students pulled out their newspapers when Max began one of his lectures. Max took it in good humor. There were times, of course, when Max waited until after the seminar to pull the aspiring scientist aside and point out that the talk was "the worst I've ever heard." Max used that phrase so often that one visitor to Caltech opened a series of lectures by stating how pleased he was to be giving more than one seminar—at least one would not be the worst Max had ever heard. Max smiled.

While Max was not always easy to be around, he made finding the most fundamental truths a team effort. His followers knew that his brusqueness came not from arrogance, but from a sincere search for understanding. He always had good suggestions, and he made it easier for his associates to be good scientists. His "I don't believe a word of it" often sent his disciples back to their lab benches for data that would force him to believe it.

Max set standards for molecular biology that lasted well into the 1970s. However, he was not an outstandingly original biologist. Even though his immediate conclusions were almost always right, the more general speculations built on these were mostly wrong. Where he went on intuition rather than logic, he was often wrong. As pointed out by one of his proteges, Max's own studies are more significant for *how* things were learned than for *what* was learned—Max brought new quantitative methods to bear on the study of microorganisms. His intellectual discipline, reinforced by that of his physicist colleagues and transmitted through his disciples, set molecular biology apart from other fields of biology.

The striking success of early molecular biology attracted many young chemists and physicists, who, armed with the Watson-Crick model of DNA, set out to answer the next set of "important" questions. The golden age of molecular biology began, and with it came the basic understanding of how DNA helps make us what we are. The major points are outlined in the next chapter.

Practical Considerations

- Atoms and molecules act according to specific rules. Our bodies are composed of molecules, so our bodies function according to specific rules. Understanding those rules gives us manipulative powers.

- DNA molecules hold the information for making the other molecules of our bodies. Tiny changes in DNA can have profound effects on body function, and understanding particular changes allows us to predict future health. Being able to make particular changes allows us to alter body fuction.

- DNA molecules differ slightly from person to person (except for identical twins). These differences can be used to track persons.

- The genetic revolution was built entirely on curiosity-driven science. Scientific breakthroughs come from support of basic, not applied, research.

The Golden Age

The Watson-Crick model of DNA provided a structural framework for asking specific questions about how genetic information is stored, how it is copied, and how it is used. Max's disciples, joined by equally tough-minded biochemists, focused their questions on the simple lives of *E. coli* and its phages. Schemes were devised for obtaining mutants with desired properties, and microorganisms were grown by the trillions to provide enough material for molecular analyses. Explorers of this unmapped world rose quickly to prominence, only to be pushed aside and usually forgotten as some newcomer brought forth an even more exciting discovery.

Arthur Kornberg is one of those explorers who has not been forgotten. In 1950, three years before the Watson-Crick model was developed, he began a concerted effort to determine precisely how cells make new DNA molecules. His approach, time honored among biochemists, was to take molecules out of living cells and then get them to perform their normal function in test tubes. He knew that life processes involve changing the structure of molecules, usually by adding, subtracting, or rearranging atoms (these changes are known as chemical reactions). Kornberg also knew that the synthesis and destruction of almost every cell part is regulated by protein molecules called enzymes (for now we can consider enzymes to be molecular workers, each specialized to speed up a particular reaction). He believed that understanding any life process required an intimate knowledge of the enzymes involved.

40

The Golden Age

Kornberg and other biochemists generally start their work by breaking cells open. Then they test the sap from the cells for its ability to carry out the particular process they wish to study. This is often done by adding specific small molecules that can be watched as they change to other molecules. The trick is to guess what molecular interactions are involved in a particular process so the correct small molecules can be added to the broken cell mixtures. Over a period of weeks or months, the biochemists gradually throw away all cellular molecules not needed for the process, leaving only the required enzymes and cofactors (helper molecules).

In the late 1940s Kornberg hunted down enzymes with a passion, uncovering one after another to define the synthesis of molecules that carry energy for driving chemical reactions. He was an enzyme collector. In some ways he was like a parent to his enzymes, always concerned for their whereabouts and safety. He couldn't leave the laboratory at night without knowing how much of each enzyme had been recovered from the day's effort. With that knowledge he could spend evenings planning the next day's experiment. Thus his days would be free for lab work, although his brand of biochemistry could hardly be called work. To him, purifying an enzyme "seemed like the ascent of an uncharted mountain: the logistics resembled supplying successively higher base camps; protein fatalities and confusing contaminants resembled the adventure of unexpected storms and hardships. Gratifying views along the way fed the anticipation of what would be seen from the top. The ultimate reward of a pure enzyme was tantamount to the unrestricted and commanding view from the summit."

Kornberg began his DNA quest by seeking the enzymes that make the building blocks of DNA, the nucleotides. Within a few years he had those enzymes in hand. He then decided to search for an enzyme that might put the nucleotides together and form DNA. He knew from the synthesis of other polymers that subunits tend to be attached to the ends of preexisting polymers. Thus he added some DNA to his reaction mixture to provide ends. He then looked for a small, radioactive nucleotide being joined to a long DNA molecule. If the polymerizing enzyme were present and active, the long DNA would become radioactive, a process he could easily detect. Kornberg also knew that his

DOUBLE-EDGED SWORD

reaction mixtures contained nucleases, enzymes that would cut up new DNA as it was being made. By adding a large amount of old DNA, he hoped to protect the new DNA from the nucleases. The old DNA also served a purpose that was not obvious beforehand—Kornberg didn't know the exact chemical form the nucleotides needed for joining, so he couldn't just add the correct ones to the mixture. As it turned out, the nucleases cut his old DNA into nucleotides that were then processed appropriately by five other enzymes in the mixture. That gave him the new materials needed for making DNA. By 1957 Kornberg had added to his collection an enzyme that could construct new DNA from nucleotides. Even though he knew about the Watson-Crick model for DNA, Kornberg was still astounded to find that his new enzyme, which he called DNA polymerase, joins nucleotides in the order dictated by those in old DNA. Indeed, his enzyme absolutely required the presence of an old DNA template. This feature provided support for the prediction that information in old DNA is used to make new DNA.

We now know that a complex of several enzymes travels along DNA, forcing the two strands apart before creating two new, complementary strands. The complex uses the nucleotide sequence of the existing strands and the complementary base-pairing rule to determine the order of nucleotides joined to make the two new strands. To reduce errors, the enzyme also proofreads the product and replaces incorrect nucleotides. Surprisingly, Kornberg's DNA polymerase, and others found later, cannot begin making DNA by themselves—they can only add nucleotides to the end of a preexisting strand. Beginning a new strand is another complex process, one called initiation of DNA replication. Eventually Kornberg and others got that to work in the test tube, too. In the 1990s we still need to discover what determines when initiation will occur, for timing of initiation of DNA replication is intimately related to the control of cell growth and perhaps to some forms of cancer.

So far I have introduced two major ideas. First, DNA is the genetic material, so it determines the properties of cells. Second, enzymes control the chemical reactions of cells, so they also determine the properties of cells. The two ideas mesh when one realizes that information in DNA determines the properties of enzymes; DNA acts through enzymes to

The Golden Age

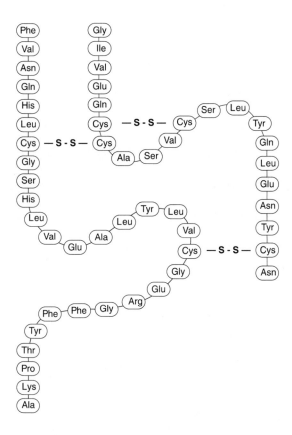

Figure 3.1. A protein as a chain of amino acids.

In the small protein called insulin, two short chains of amino acids (abbreviated by three-letter acronyms) are hooked together by sulfur atoms (labeled S). The three-dimensional folding of these chains is not shown. Figure adapted from *Biological Science* by W. Keeton, W.W. Norton and Co., New York.

guide cell activities. Obviously, understanding DNA action and genetic disease requires knowledge of enzymes and other proteins.

Proteins are chainlike molecules, often containing hundreds to thousands of links (Figure 3.1). Each link is a substance known as an amino

DOUBLE-EDGED SWORD

acid, of which there are twenty different types. Some amino acids in a protein attract each other, while others repel one another. These attracting and repelling forces cause the protein to fold in a specific way (see example in Figure 3.2). The specific folding then enables the protein to facilitate particular chemical reactions or to act as a structural component of a virus or cell.

Enzymes comprise a special class of protein that functions in cells like a child's fingers while building or destroying particular structures with construction toys. For example, the fingers put particular pieces together to form walls, floors, and doors. If free building blocks are used up, structures made yesterday can be torn apart to allow today's construction to continue. Compared with enzymes, a child's fingers are very versatile,

Figure 3.2. Protein folding.

The protein called myoglobin is composed of a single chain of 151 amino acids that fold into regions where the amino acids form a helix (spiral). Nonhelical regions occur between the helices and allow the folding. Figure adapted from *Biological Science* by W. Keeton, W.W. Norton and Co., New York.

The Golden Age

for the same fingers can work on pieces of many different sizes and shapes. In contrast, each enzyme type is highly specialized, and each facilitates only one or a few closely related conversions of one molecule to another. Thus a cell must have thousands of different enzymes to build its parts. But enzymes are much faster than fingers: some can stimulate ten billion conversions per minute. Often an enzyme surface contains a cavity that fits only a particular type of small molecule. There the small molecule sits and temporarily interacts with the atoms of the enzyme cavity. Such interactions can weaken bonds between specific atoms in the small molecule and cause the small molecule to break. In some cases, the interaction allows the small molecule to join its atoms to another small molecule or to the end of a large one brought close through its attachment to the enzyme.

The understanding of protein structure and action allowed molecular biologists to address three more questions: How is information for protein structure carried by DNA? How is the information used to make specific proteins? And how is the timing of specific protein production controlled? Watson and Crick helped lead the growing pack of molecular biologists toward the answers to these questions.

We now know that information is held in DNA much as we find information in movie film. Both film and DNA are very long strands composed of large information units. In film we call those units scenes, and in DNA we call them genes. The scenes are further divided into frames, which would correspond to the individual nucleotides of DNA. Both scenes and genes have distinct frames/nucleotides that specify beginnings and ends. Moreover, a piece of film can be cut at any frame and attached to another piece of film. Likewise, a piece of DNA can be cut at any nucleotide and joined to another piece of DNA.

Some aspects of DNA are easier to understand if we consider DNA information to be arranged like a written sentence. The genetic letters, A, G, C, and T, are functionally read as words. The words, also called codons, make up genes, which are hundreds to thousands of codons long. In this analogy, a gene is equivalent to a DNA sentence, one in which there are no spaces between the genetic words (Figure 3.3). To get the alignment right, the gene starts at a precise nucleotide—if you

DOUBLE-EDGED SWORD

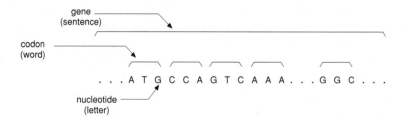

Figure 3.3. Information in DNA.

Information in DNA is arranged linearly, much in the way we write. Chemical structures called nucleotides (A, G, T, and C) are the letters of the text. They are arranged into three-letter words called codons, which make up genes. Genes are equivalent to sentences.

put an extra nucleotide into the middle of a gene, the information from that point to the end of the gene would make no sense (Figure 3.4b).

In 1961 Crick was working with the idea that chemicals resembling bases could slip into DNA and cause the replication machinery to either add or delete a base. That would create a mutation. To obtain a collection of phage mutants, he would add one of the baselike chemicals to bacteria infected with phages and then look for phage mutants that could grow in one type of bacterium but not in another (the normal phages could grow in both types of bacterium). From other studies, Crick knew that such mutants had defects in a particular gene. When he infected bacteria with two phages that each had a different mutation in the gene, he obtained phage recombinants, phages resulting from DNA molecules that had been broken and rejoined. The recombinants behaved as if they had no mutation, although in reality they had two mutations. It was as if each mutation corrected for the presence of the other (Figure 3.4c). He soon discovered that some pairs of mutations were corrective, while others were not. He was clearly onto something new, and he pressed hard with the work.

Crick found that if he set up an experiment in the morning, he could get the result by late afternoon. He needed only a few minutes to analyze the experiment, and that gave him time to start a second experiment in

a. correct	JOE SAW THE BOY HIT THE CAR AND THE DOG
b. insertion of 1 X:	JOE SAX WTH EBO YHI TTH ECA RAN DTH EDO G
c. insertion of X, deletion of H:	JOE SAX WTE BOY HIT THE CAR AND THE DOG
d. insertion of 2 X's:	JOE SAX WTH XEB OYH ITT HEC ARA NDT HED OG
e. insertion of 3 X's:	JOE XSA WXT HEB OYX HIT THE CAR AND THE DOG

Figure 3.4. Shift in reading frame of a gene.

Insertion of the letter X in the sentence, or an extra nucleotide in a gene, shifts the reading frame so all downstream information makes no sense. Likewise, insertion of two Xs creates nonsense. Insertion of one X shifts the reading frame out of register, but deletion of a letter elsewhere puts it back in the proper reading frame. Insertion of three Xs disturbs a portion of the gene, but the proper reading frame is restored. In some cases, the three insertions also restore gene function.

the evening. The second experiment would be completed during the night, so by morning he was ready to start two more experiments. Just as Delbrück had discovered two decades earlier, phage again seemed to be God's gift to physicists-turned-biologists.

Soon Crick had divided his mutants into two categories, + and −. Phage containing two mutations, with one from each category, behaved like normal phage (Figure 3.4c), while those with mutations from either + or − were defective (Figure 3.4d). He realized that he was in a position to determine how many nucleotides make up a genetic word, a codon. One afternoon he set up a phage infection to produce a triple mutant in which all three mutations were of the same class. If that mutant behaved like a normal phage, the code must be read in threes (Figure 3.4e). That evening held a rare moment of discovery: when Crick and a colleague returned to the lab, they saw normal plaques arising from the triple mutant. For a few minutes they were the only persons in the world who knew that the genetic code is read as triplets.

Subsequent experiments showed that a gene determines the molecular structure of a particular protein through a correspondence between codons in DNA and amino acids in protein. This correspondence is called the genetic code, and by the early 1960s the code was broken: *particular* nucleotide triplets in DNA were shown to correspond to *particular* amino acids in proteins. For example, the codon ATG specifies the amino acid called methionine, and the positions of ATG codons in a gene establish where methionines will be placed in the protein chain encoded by the gene. Since the order and number of nucleotides varies from gene to gene, different proteins have different numbers and orders of amino acids. And since the order of amino acids in a protein is responsible for how the protein folds, the order of the amino acids, and ultimately the order of nucleotides in DNA, determines how the protein will act as a part of a virus or cell. That is how DNA controls biological chemistry.

The relationship between nucleotide sequence in DNA and amino acid sequence in protein makes it easy to understand mutations: they are changes in the nucleotide sequence of DNA that alter the order of amino acids in a particular protein type. That amino acid alteration changes the ability of the protein to do its predetermined job. Cystic fibrosis, for example, is often due to the loss of a particular nucleotide triplet that causes a chloride channel protein to lack the amino acid phenylalanine at amino acid position number 507. If your DNA lacks that triplet in the chloride channel gene, you either have cystic fibrosis or are a carrier of the disease (occasionally the symptoms are mild, as if another gene sometimes modulates the severity of the disease).

Once it was clear that the order of nucleotides in DNA corresponds to the order of amino acids in proteins, it became possible to ask how proteins are made from DNA information. The process is called gene expression. By the early 1960s it was understood that proteins are made by a two-step process in which the information is carried from the storage molecule (DNA) to workbenches (ribosomes) where the information is used to make proteins. In the first step, which is called transcription, the information contained in a specific gene is copied by making RNA molecules having the same nucleotide sequence as the gene (Figure 3.5). RNA is similar to DNA, but it is shorter, generally single stranded instead of double stranded, and has subunits (nucleotides) that differ slightly from those in DNA. In addition, RNA contains the base uracil, abbreviated by the letter U, instead of the base thymine (T) found in DNA (the nucleotides in RNA are abbreviated A, G, C, and U). Transcription is much like making a videotape of a scene in a motion picture film: the resulting information is the same, but the form is different. In this case, the tape is much shorter than the film.

The second step of gene expression involves the conversion of information from RNA to protein. This step is called translation because information specified by a four-letter alphabet, the four different nucleotides, is respecified by a twenty-letter alphabet (the twenty different amino acids). The RNA gene copies, which are called messenger RNA molecules, move to ribosomes, large subcellular RNA-protein complexes where proteins are made. A ribosome attaches near one end of a messenger RNA and begins to move along it, pausing momentarily at each codon. At each pause, an amino acid of the type specified by that codon aligns on the ribosome and becomes bonded to the amino acid that immediately preceded it. The ribosome continues down the messenger RNA, aligning and joining amino acids in the order specified by the codons of the messenger RNA. Eventually the ribosome reaches a codon that signals the ribosome to stop. Then the new protein separates from the ribosome, travels to its place of work in the cell, and begins doing its job.

The general understanding of DNA, RNA, and protein synthesis brought the golden age of molecular biology to a close in the late 1960s. In 1969 Max, Luria, and Hershey shared the Nobel prize, and Stent, always one of the more eloquent of Max's disciples, wrote a book about the golden age. The prize and the book seemed to end the era officially. Of course, discoveries didn't stop pouring in, but they were considered to be mere details. Many of Max's followers felt that all the important questions had been answered (as mentioned earlier, Max himself had left the scene long before, realizing in 1950 that no new physical principles would emerge from gene studies). To drive the point home, Stent delivered a series of public lectures describing the end of the "important" gene questions. Scientists who were students at the time seriously wondered whether their careers would be relegated to clean-up work. And they didn't wonder alone; advisors suggested that students get secure teaching jobs, since research funding would dry up when the important questions disappeared.

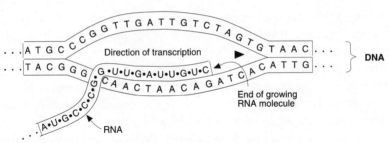

Figure 3.5.   Transcription.

During transcription, the enzyme complex called RNA polymerase causes DNA strands to separate over a short region. The polymerase moves along the DNA, and as it does, it forms an RNA chain using free nucleotides. The order of the nucleotides in RNA is determined by the order of nucleotides in one of the DNA strands by the complementary base-pairing rule (that rule states that A always pairs with U or T, and G always pairs with C).

The process of transcription allows the cell to selectively access information in DNA. That in turn makes it possible for cells to use information from some genes more than from others. By controlling when transcription of a particular gene occurs, the cell gains control over the timing of gene utilization. For example, we have in our DNA several different genes for the oxygen-carrying blood protein called hemoglobin. At different stages in our lives (embryo, fetus, adult) our bodies require a different form of hemoglobin; consequently, the different genes are used at different times. This type of understanding could be useful, since in principle it provides us with a way to combat some forms of thalassemia, a genetic blood disease in which an adult hemoglobin gene is defective: we just need to figure out a way to cause hemoglobin to be made from a fetal gene that is normally shut off shortly after birth. The fetal gene might not be quite as good as the adult one, but a partially active protein is better than none.

The Golden Age

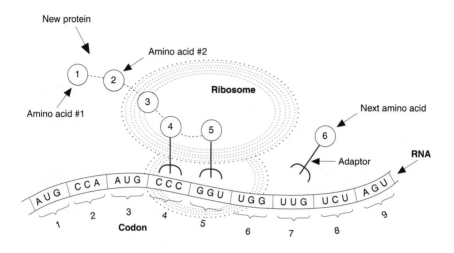

Figure 3.6. Translation.

RNA binds to a ribosome, which then moves along the RNA and uses the information in the RNA to align and join amino acids into a chain that folds to form the finished protein. The correct amino acid is aligned by an adaptor RNA molecule (transfer RNA) that recognizes both a specfic RNA codon and a specific amino acid.

Then came gene cloning. Five years after Stent's pronouncement, Paul Berg and his colleagues at Stanford University began cutting genes out of one type of DNA and inserting them into other DNA molecules. This technique, which used bacterial cells to help purify small bits of DNA, made detailed studies of genes possible. During the mid-1970s methods became available for determining the nucleotide sequence of large regions of DNA. Then biologists from all fields began using DNA sequences to make new combinations of DNA, laying the groundwork for an explosion of applications. Nobel prizes continued to be awarded to molecular biologists.

By the 1990s the mental discipline Max had given the field had diluted considerably. Molecular anatomists began collecting DNA sequence information as a resource for future, undefined experiments, and projects took on a more applied tone as national priorities changed, first to address AIDS and then tuberculosis and breast cancer. The age of the gene hunters had begun.

———+·

Before considering applications of molecular biology, it is useful to recap the principles derived from phage and bacterial studies that apply to our own bodies. The main theme is that all life forms are masses of minute molecules dutifully obeying the physical rules that govern all atoms. Many of our molecules behave as though they've been programmed for specific jobs. Some align with each other to form thin membranes, others package together to make long tubules or tough protective coverings, and still others fold many, many times into compact forms that provide organization for billions of bits of genetic information. Dynamic interactions occur. For example, some enzymes can lock onto a DNA molecule, distort the bonds between the nucleotides, and break the DNA. Other enzymes are builders, bringing small molecules together and joining them to form giant molecules. And once built, many molecule types travel from one cellular location to another and from one body part to another.

To perform their duties, molecules must recognize each other. Recognition occurs by tiny contours on molecular surfaces making lock-and-key-type connections. The individual connections are frequently very weak; consequently, when two molecules do lock together, it is because many tiny connectors align perfectly. If the fit is not exactly right, the molecules don't stick when they bump together. This principle also applies to different regions of the same molecule fitting together to give the proper fold. Incorrect amino acids in proteins affect these fits, and so they result in genetic disease.

Unlike the cells of microorganisms, which are often independent and free-living, our cells cluster to form larger units. Molecules on cell surfaces specify the rules for cell association. These molecules create uneven surfaces of molecular protrusions and invaginations (indentations) that are capable of the lock-and-key couplings responsible for cellular recognition and joining. Such interactions are very specific, allowing only certain cell types to attach and form tissues. Specific tissues in turn mass together to form organs. Again, the structure of an organ is determined by the structure and reactivity of the molecules of the tissues that comprise it.

One of the practical consequences of molecular recognition is rejection of organ transplants. Organs from different persons often have slightly different surface molecules, and those differences can prevent one body from accepting an organ transplant from another. By studying the cell surface molecules, doctors can sort through a collection of organs to find ones most likely to work in any particular person.

The coordinated action of the molecules, cells, tissues, and organs makes our bodies function. Ultimately, molecular changes mediate coordination at all levels. For example, an environmental stimulus can cause one cellular group to secrete hormone molecules that travel through the blood and cause a second group of cells to change its molecular composition. The coordination extends even to our behavior, since there are genetic diseases that have distinctive behavior patterns associated with them. An example is Tourette syndrome, which is characterized as a collection of uncontrolled tics, twitches, belches, and obscene outbursts. Another is Huntington disease, which is discussed in the next chapter. How extensively genes influence more acceptable behavior patterns is a question now being investigated by molecular biologists. Already we've seen controversial suggestions that a variety of behaviors, most recently male homosexuality, have a genetic basis. While many of the assertions about behavior have not withstood close scientific scrutiny, we will inevitably find that many aspects of behavior do begin in the genes.

These general ideas about biological molecules, plus the known role of DNA, emphasize the great power we can attain by manipulating genes. However, the few examples mentioned in this book represent only the barest beginning, since we have about 100,000 genes to understand. Moreover, our different cell types each have their own individual constellation of proteins, with no cell containing every type of protein (the number of different protein types in a cell is on the order of thousands, with some proteins being present in only a few copies, others in thousands of copies). Cells obtain different protein compositions by selective use of information in DNA, which, with few exceptions, is the same in our trillion or so cells. This leaves us with many genes to understand and many gene control circuits to dissect.

While genes and their interacting circuits are being sorted out, one of our tasks is to protect our DNA molecules from damage. Environmental factors, such as sunlight, X-rays, and harsh chemicals, can produce small but serious alterations (mutations) in the nucleotide sequence of DNA. If these mutations occur in our germ cells (sperm and eggs), they are passed to our children at conception. When mutations arise in our somatic (body) cells, they may contribute to the development of certain types of cancer. There is currently little good advice for avoiding mutational problems. However, individuals in families prone to cancer can benefit from genetic screening programs. For example, if you inherited a defective gene that in the normal state suppresses cancerous growth, you can have frequent checkups, undergo preventive surgery, and guard against exposure to DNA-damaging agents. At the same time, you must decide who else should know about your predisposition. Genetic privacy has become a major issue for each of us as DNA methods uncover disorders long before they become visible. The privacy issue is illustrated in the next chapter through a consideration of Huntington disease, one of the most insidious of the heritable maladies.

Practical Considerations

- Information in DNA indirectly controls cell chemistry by determining protein structure. Thus cell chemistry and cell behavior can be changed permanently by altering DNA, the repository of genetic information. This is the basis of gene therapy and genetic testing.

- Proteins directly control cell chemistry by accelerating specific chemical reactions. Cell chemistry can be changed temporarily by injecting proteins. An example is the injection of insulin to control diabetes.

- Conversion of information from old DNA to new DNA or to new protein involves a number of complex steps. The molecules involved are slightly different in bacteria and human beings. That has made it possible to find a variety of antibiotics that selectively interfere with the processes in bacteria and kill them.

- The function of many genes is influenced by other genes, usually through interactions among their protein products or through the protein product of one gene influencing the expression of another gene. These control circuits among genes are so complex that we are still far from understanding human genetics. Environmental conditions also affect gene expression and make understanding genetics even more difficult.

$\dagger$

The Enemy Within

Along the shores of Lake Maricaibo lie several remote fishing villages bordered by swampy jungle. The drafty little houses, some perched high above the water on stilts, are well suited to the warm Venezuelan climate. Life there seems to flow without much hustle or bustle. But underneath a calm facade the people fear a terrible fate. Each person waits and watches for the first signs: irritability, short memory lapses, tiny jerking movements. Each knows that these small signs eventually magnify to the constant, uncontrollable spasms of Huntington disease. A doctor, writing more than seventy years ago about a victim in another part of the world, described the symptoms most poignantly:

> The door opened in an irresolute way, and an arm was thrust through with a spasmodic jerk; then a leg followed with a like unsteadiness; while the right hand twitched at the handle, the corresponding body clinging in the door edged itself round with a random attempt to close it. Then with a stagger the patient lurched forwards to our table . . . a tripping, staggering gait, hastening and stopping until he reached the chair. Yet even in the chair, you see as we did, that the man is still in incessant restless motion, like a marionette; now jerking an arm, now the trunk of the body, now shrugging a shoulder . . .

These spasms, often beginning in middle age, can last for fifteen to twenty years before lethal respiratory problems arise or the heart wears out from the constant motion. Nothing short of death halts the progression of this relentless, devastating disease.

As is common among geographically isolated groups, the people of the water villages are related, and that relatedness brought them into a gene hunt. In the mid-1980s the family numbered some 3,000 living members, with about 100 showing obvious signs of Huntington disease. Roughly 1,100 others, mostly children and young adults, had an afflicted parent; that gave them a significant chance of succumbing to El Mal, their name for the disease.

El Mal did not always thrive around Lake Maricaibo. Local legend blames the illness on sailors from Europe, and Europeans did indeed spread Huntington disease to many parts of the world. In the seventeenth century, many Europeans believed that the devil possessed individuals stricken with the disease. Consequently, communities periodically purged themselves of afflicted families, who frequently fled to new lands. Although the Venezuelan genealogy is too vague for us to identify the responsible sailor, the disease is so distinct that it is often easily traced. For example, three persons who sailed aboard the John Winthrop fleet from England to America in 1630 have been identified as carriers of the disease. They had hoped to leave the curse behind them, but of course they couldn't. Instead, they founded an American family of Huntington victims. At least three women of the family were subsequently executed as witches, one of whom became notorious as the Groton witch of 1671.

The "dancing madness" took on George Huntington's name because in 1872 he wrote an especially careful and perceptive analysis of the hereditary nature of the disease. From Huntington's work and subsequent family analyses, two clear statements emerged about the disease. First, if one parent has the disease, on average half of the children will also exhibit it when they become adults. Second, the disease appears only if at least one parent is afflicted—it never skips a generation and then comes back (sometimes a family member dies from other causes before being stricken by the disease, so the unbroken line may not be obvious). These genetic characteristics support the simple idea that the disease arises even when only one of the two copies of the Huntington gene is defective (chromosomes come in pairs, so there are two copies

DOUBLE-EDGED SWORD

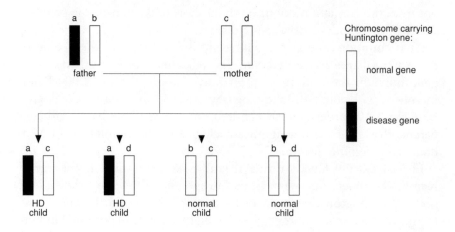

Figure 4.1. Pattern of inheritance of Huntington disease.

of each gene). Geneticists call this type of condition dominant: if the diseased gene copy is present at all, the carrier will eventually be stricken (see Figure 4.1). This contrasts with cystic fibrosis, in which both copies of the responsible gene need to be defective for the disease to occur (see Figure 1.1).

The Venezuelan villagers are trapped in the genetic cycle of El Mal; we can offer them neither therapies nor immunization programs. With meager resources and many small children to feed, one family cluster after another crumbles as parents succumb. Since Huntington disease is relatively rare (there are about 30,000 cases in the United States), afflicted persons tend to have one normal and one disease-producing copy of the chromosome bearing the responsible gene. But even with only one parent carrying one copy of the disease gene per cell, each child still has a 50–50 chance of getting the disease (each child inherits either a good gene or a bad gene from the afflicted parent). Every descendant, uncertain of which gene he or she has, approaches middle age with great foreboding, dreading the onset of symptoms and hoping not to have passed the curse to the next generation.

A second thread of the story begins in southern California. In the late 1960s, a successful psychoanalyst, Milton Wexler, noticed his wife behaving strangely. He suspected Huntington disease, since her brothers had suffered from it years earlier. At some point he would have to inform his two daughters, for they could be next. The young women were then in their early twenties, and the news would surely devastate them. But they needed the opportunity to focus their lives quickly. Soon after the diagnosis was official, the psychoanalyst told his daughters that they could end up like their mother. They were stunned.

Of course, each daughter had only a 50–50 chance of being afflicted. That allowed for some hope, but the weight hanging over them was so great that they quickly abandoned the idea of having children of their own. They wouldn't risk condemning anyone to the torture of this disease.

One of the daughters, Nancy, pursued a career in experimental psychology. She could empathize with the mental agony of those at risk for genetic disease as few other psychologists could, and she gradually moved closer and closer to the study of her own family demon, Huntington disease. Her father also became active in the Huntington arena by raising research funds. He hoped that a solution would come from medical research, and he wanted to speed it along so his daughters might be spared.

Years went by without progress. Biochemists and geneticists tracked members of families with Huntington disease, but the search failed to uncover the chemical defect. Nancy Wexler and her sister were stuck with the same prospects as the Venezuelans, constantly wondering if the latest dropped glass or coffee spill signaled the onset of madness.

A third part of the story centers on tracking DNA molecules. By the late 1970s biologists realized that if they could identify a region of DNA, a gene, responsible for a particular genetic disease, then analysis of the gene might provide insights into the disease process or at least generate a diagnostic test. (In Chapter 1 we saw how the application of nucleotide sequence information was used in a fetal test for cystic fibrosis.) The initial problem was to locate the Huntington disease gene among the 100,000 other human genes and tease it free from them for biochemical studies.

DOUBLE-EDGED SWORD

At the time, Arthur Kornberg and other DNA biochemists had provided molecular biologists with a small collection of enzymes for cutting and joining DNA molecules. The cutters were called restriction endonucleases because they restricted (limited) the ability of phages to infect bacteria. These enzymes serve as a natural defense for the bacteria by cutting foreign DNA when it enters a cell. These nucleases are important as engineering tools because they cut at specific sites on DNA (at specific nucleotide sequences). That generates DNA fragments, restriction fragments, having discrete lengths (Figure 4.2).

Curiosity led biologists to examine the nucleotide sequence in restriction fragments of DNA obtained from a large number of people. For most sequences examined, people are very similar. There are, however, specific regions of DNA that vary considerably from one person to another. Some of these differences change the spacing between the places where a restriction endonuclease cuts the DNA. This results in a particular restriction fragment whose length can vary from person to person (Figure 4.3). Regions of DNA in which restriction fragment length varies are called restriction fragment length polymorphisms (*polymorphism* means "multiple forms"; the acronym for such a region is RFLP, often pronounced as "riflip"). In the motion picture analogy used in earlier chapters, a polymorphic region would correspond to a commercial inserted into a film shown on television. For a given commercial time slot there could be a variety of nearly identical ads that would differ from one local channel to another. The variation in ads at each commercial interruption would make the same film differ slightly from channel to channel, just as DNA differs from person to person.

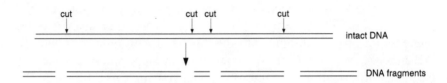

Figure 4.2.   Cuts at specific locations produce DNA fragments having discrete sizes.

The Enemy Within

Figure 4.3. A polymorphic region of DNA.

In a specific region of DNA the distance between two particular restriction sites (arrows) vary among the four persons illustrated. Thus when DNA is cut with the enzyme, the sizes of the pieces generated will differ from one person to the next.

Biologists soon realized that they could track the inheritance of a *particular* restriction fragment length from a polymorphic region just as they could trace inherited features such as eye color or Huntington disease. To do this, DNA molecules extracted from parents and children were cut with a restriction endonuclease. Then the fragment length from the specific polymorphic region was measured for each DNA sample. The fragment lengths were the same for parent and child under conditions where inheritance was expected—molecular methods could be used to follow a short region of DNA from one generation to the next.

The discovery that many different polymorphic regions exist and that individual fragment lengths are inherited like genes opened the way for a new breed of biologists, the gene hunters. Armed with the enzyme tools discovered by Kornberg and the enzyme hunters, the gene hunters set out to locate the polymorphic regions present in human DNA and create a road map for finding genes. As detailed below, a particular heritable trait would be located near a specific polymorphic region, the gene responsible for the trait would be joined to small infectious DNA, and the gene would be amplified in bacteria. Then the DNA would be purified so the gene hunters could add the gene to their collections.

Part of the gene hunting strategy involved recombination, a natural

DNA breaking-rejoining phenomenon that has been discussed earlier as a part of phage biology (see Figure 2.1). Recombination also occurs in human beings where it keeps children from being too much like their parents. Our chromosomes, which contain single pieces of DNA, are inherited as discrete units; thus two characteristics resulting from genes located on the same chromosome are also inherited as a unit (in Figure 4.4, Huntington disease would be inherited with the feature labeled *A*). Such characteristics are said to be linked, just as scenes in a movie are linked. When sperm and eggs are made, the two copies of a chromosomal pair align, break, and exchange portions (in Figure 4.4 notice how darkened region A switches from DNA #1 to DNA #2). This results in a feature found on one chromosome, such as the defect that causes Huntington disease, being switched to the other member of the chromosome pair. Then Huntington disease is inherited with the feature labeled *C* in Figure 4.4 instead of the one labeled *A*. This switching process is called genetic recombination.

The recombinational breaks in DNA occur more or less in random locations, so the chance of a break occurring between two particular genes increases as the distance between the genes increases (in Figure 4.4, this distance is represented by the stippled region). Thus the distance between two genetic characteristics can be estimated by how often their linked inheritance is separated by recombination. Once the distance between genes is known, the genes can be arranged on a linear map.

Since recombination can affect the accuracy of a genetic test, it is worth illustrating the concept. Suppose children in a particular family tend to inherit either long angular noses along with large ears or small rounded noses along with small ears (the genes for nose shape and ear size linked on the same chromosome). Occasionally a child might be found who has a small rounded nose and large ears. This would result from recombination occurring between the nose shape gene and the ear size gene. How often this occurs relative to the standard angular nose-large ear and rounded nose-small ear inheritance pattern would be a crude measure of the distance between the two genes.

When different polymorphic (variable) regions of DNA are mapped instead of nose shape and ear size, DNA restriction fragment lengths

The Enemy Within

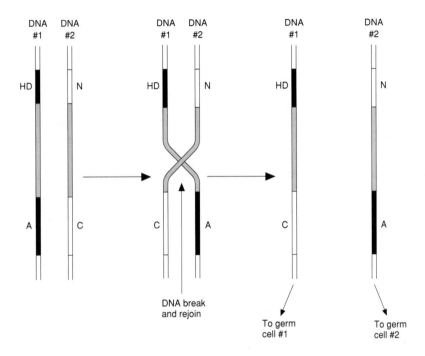

Figure 4.4. Separation of inheritance of two traits by recombination during germ cell formation.

HD and N represent the disease and normal form of the Huntington disease gene, respectively. A and C are two forms of a nearby RFLP called G8.

rather than noses and ears are examined. Building a DNA map with restriction fragments is simpler than with genes because polymorphic regions are equivalent to single gene features while characteristics such as nose shape and ear size are often complicated by dominance factors and the action of several genes. Once the relative positions of the polymorphic regions are determined, the regions can serve as genetic signposts, much like benchmarks used by surveyors. For example, with the television movie idea, a particular scene (gene) can be more easily located if you can quickly scan to a commercial interruption (polymorphic region) known to be near the scene (gene).

The different polymorphic regions are identified by a biochemical test called Southern blotting (described in Chapter 8). Thus, if a polymorphic region could be found on DNA close to the Huntington disease gene, the polymorphic region could be followed from one generation to another and become a predictive marker of disease. Moreover, the region could be used to identify and purify pieces of DNA that contain the disease gene, and that would eventually lead to nucleotide sequence information and perhaps insights leading to a treatment. This general idea, which could be applied to any disease gene, stimulated a rush to find polymorphic regions.

In 1979 biologists thought that they might need a decade to turn up a polymorphic region close to the Huntington gene, since the odds for any particular polymorphic region being near the gene were very small. Before joining the polymorphism collectors, James Gusella at Harvard University decided to examine the few regions already available. He hoped to find a polymorphic region that had a distinctive restriction fragment length always inherited with Huntington disease. For this, he needed to check DNA samples from a large family in which many members had the disease.

Nancy Wexler, the psychoanalyst's daughter, brought the Venezuelans and molecular biologists together. When she learned of the interrelated Venezuelan community in the late 1970s, she didn't know much about DNA polymorphisms. But she did recognize the general value of such a large family for the study of genetic disease. Initially, she thought that a family analysis would help her find an afflicted person having both copies of the Huntington gene in the disease form. Then a physical examination of such a person might give insight into the disease process. Such a strategy had worked well with a cholesterol disease. In 1979 she trekked to the water villages of Lake Maricaibo. There she found a husband and wife who each had El Mal. Statistically, a quarter of their thirteen children should have had both gene copies in the disease form (the inheritance pattern would be the same as that shown in Figure 1.1). None showed symptoms, and at the time there was no way for Wexler to determine the gene composition of any particular child. Moreover, the couple had suffered two miscarriages. Perhaps the miscarried fetuses were the ones who had two bad copies of the gene; perhaps such babies never make it to birth. Regardless of the explanation, Wexler made little progress along this line of study (she learned much later that the symptoms are the same for persons with one and two copies of the disease gene). This visit, however, did build a rapport between her and the large Venezuelan community.

When Wexler returned to the United States, she learned about the search for a polymorphic region always inherited with Huntington disease. From Venezuela she could get the many family histories and blood samples needed to look for the right region. With two more expeditions to Venezuela, she obtained 570 blood samples along with a detailed family tree. Meanwhile, the biologists had started pilot studies with a small American family from Iowa. By 1982 they had found a polymorphic region called G8 that seemed to be traveling with the disease from one generation to the next in the Iowa family. Examination of the huge Venezuelan collection firmly established that G8 DNA fragments from Huntington victims and from unaffected family members had different lengths—by an incredible stroke of luck Wexler and the biologists had found a DNA marker for the disease after looking at only a handful of different polymorphic regions. They had found the proverbial needle in the haystack.

The G8 region has several forms in human populations. The early studies revealed four forms—that is, four restriction fragment lengths. For convenience, the forms were called G8-A, G8-B, G8-C, and G8-D. In the Venezuelan family, the G8-C form is inherited with the disease, and it isn't found on normal chromosomes. In the American family, it's the G8-A form that is inherited with the disease; the G8-B, G8-C, and G8-D forms are associated only with normal chromosomes.

Once the association of Huntington disease with G8 was made, G8 could be used to predict whether an unborn child is doomed to suffer from Huntington disease fifty years hence. To look for the presence of the Huntington disease gene, the first step was to analyze the G8 region of DNA from family members, usually from three generations. This revealed which form, A, B, C, or D, was carried by those afflicted with the disease in that particular family. Then DNA from the test subject was examined to see whether he or she had the form of G8 associated with the disease. Nothing was known about the Huntington gene itself except that it mapped near a polymorphic region (in this case G8) whose cellular function, if any, was also unknown.

The G8 test was imperfect, since a positive result meant that the person had only a 95 percent chance of getting Huntington disease. The odds were not 100 percent because recombination occasionally occurs between the Huntington gene and the G8 marker, causing the two to be inherited separately (see Figure 4.4). In the Venezuelan family, a case was seen in which the disease switched from an association with G8-C to G8-A. The imperfection in the G8 test also affected those diagnosed as safe from Huntington disease: they still had a 5 percent chance of getting the disease due to recombination before they were conceived. Subsequently, other polymorphic regions closer to the gene were found; that reduced the chance of recombination and increased the accuracy of the test.

At first Wexler and her sister were euphoric over how rapidly a test for Huntington disease had emerged. Then test results began to come in for other people. For some persons the news was good; for others it was bad. It quickly became apparent that the bad news was more than just bad: it removed all hope for a normal life. It became clear that not everyone at risk should take a test for a disease that has no cure.

Our story is without resolution. Those at risk still live as though their homes are at the base of a cliff, just waiting for an avalanche to crush them. Wexler has now placed more than 10,000 entries on the Venezuelan family tree, collected some 2,000 blood samples, headed government advisory committees concerning genetic testing, and won the Albert Lasker Award (one of the highest honors in American medicine). By early 1993 her efforts contributed to identification of a gene thought to be responsible for Huntington disease. Nucleotide sequence analysis revealed a repeating nucleotide triplet (CAG) in the gene. There are from eleven to thirty-four repeats in normal chromosomes and from forty-two to over sixty-six in Huntington chromosomes. This region of repeats seems to be unstable in the disease form of the gene, with higher numbers of repeats correlating with earlier onset of disease symptoms.

Exactly how a high number of CAG repeats leads to disease is not yet known, but its correlation with disease should make the test for Huntington disease much more accurate and should eliminate the need for testing others in the family.

Whether Wexler and her sister have been tested is a private matter, as it should be for anyone. However, maintaining privacy is not always easy, and as pointed out below, revealing genetic predisposition to disease can have serious consequences.

———+··

We share a basic body plan that nature has been refining for millions of years. But each of us also has a unique genetic history, since DNA molecules acquire changes as they pass from generation to generation. Some of these changes occur rarely. When they persist in a family, they give it characteristic features. If you look into a mirror and compare your image with pictures of your own parents, grandparents, and even great-grandparents, you may recognize traces of your genetic history. With DNA analysis and Huntington disease, we see how future characteristics can be made visible. Knowing what is written in our genes can unlock a new set of opportunities, particularly with respect to family planning. But it also opens up new problems: for the afflicted, genetic information can be emotionally and economically devastating.

Our consideration of Huntington disease makes it clear why protecting genetic privacy is important: families can be branded forever. At one time, victims of this disease were considered demonic. To escape persecution at the stake, family members sometimes changed their surnames, emigrated to new countries, and severed all ties with relatives. More recently, persons admitting to a family history of Huntington disease have been denied admission to certain high-risk military programs and to professional schools requiring long training periods. The combined power of DNA testing and computerized data banks will make it increasingly difficult for a family to escape the stigma and economic penalties that genetic disease can impose.

Organized screening programs need not be in place for family "branding" to occur. For example, genetic branding can come from well-intentioned research projects. Even a project as seemingly benign as examining preserved tissue of Abraham Lincoln has potentially far-reaching implications. It has been suggested that Lincoln suffered from Marfan syndrome, a genetic disease that causes, among other things, weakness in blood vessels. The Marfan Foundation has argued that a positive test result for the syndrome would show afflicted persons that they can lead very productive lives in spite of the disease. But such a DNA result might also present anyone related to Lincoln with a new set of medically related insurance or employment problems, since the case has gained considerable publicity. In this particular situation, there are apparently no living descendants of Lincoln to whom he could have passed the disease, and a panel of experts concluded that there are no compelling legal or ethical reasons to prevent experiments along these lines. But how did Lincoln get the Marfan gene? Might the gene also occur in more distant branches of the Lincoln family tree, and might individuals in those branches someday be identified by insurance companies as big risks?

We see another example with a cluster of Jews that recently emigrated to Israel from Yemen. Among them, about one in 5,000 babies (two to three times the normal frequency) suffer from the recessive genetic disease called phenylketonuria (PKU). These babies lack an enzyme called phenylalanine hydroxylase, which breaks apart the amino acid phenylalanine. Without the enzyme, phenylalanine and some of its by-products accumulate in the bloodstream and cause a variety of symptoms, including mental retardation. Medical geneticists wondered whether the Yemenites constitute a group of related families. Initial studies suggested that the individual families had come from twenty-five different towns scattered throughout Yemen and were, for all practical purposes, unrelated. But further research into family history told a different story. Until about 200 years ago, each family had ancestors who lived in the city of San'a. Since the Yemenite Jews were close-knit, as might be expected of such a group living in a Moslem country, there was little marriage outside the group. That could explain the high incidence of PKU. Perhaps hundreds of years ago, an individual was born with a nucleotide change in one of the two copies of the phenylalanine hydroxylase gene. The baby probably grew up and lived with no ill effects, since each human being has two copies of this gene and only one good copy is needed for normal function. Thus the genetic alteration could have passed down unnoticed from generation to generation. Occasionally, people married who each carried an altered gene. A quarter of their babies would receive a defective gene from both parents and would suffer from PKU (the calculation is the same as that used for cystic fibrosis; see Figure 1.1). Today as many as one in forty Yemenite Jews carries the genetic defect. Family members can now be alerted to the problem and prepared for the special diet that a PKU baby needs. But in the process, the family has been tagged with the PKU label.

The PKU brand is not nearly as devastating as the Huntington label, since avoiding certain foods and particular artificial sweeteners can alleviate disease symptoms. Indeed, treating PKU has been one of the genetic success stories, and PKU screening of newborns is now standard practice in developed nations. But such widespread screening without serious privacy safeguards raises issues about future discriminatory use of the information.

Branding can also occur when a family itself engages in amateur genetics. An example is the discovery of Joseph disease, a syndrome resembling Huntington disease. About twenty-five years ago, a California woman looked for a pattern among the members of her family who had been striken with this degenerative neurological disorder. She interviewed hundreds of relatives and eventually traced her family curse back to a Portuguese sailor. Her family tree made the hereditary trail of the disease obvious. Many undiagnosed and unsuspecting family members were shocked to learn that they might soon be afflicted.

Inherited disease sometimes emerges suddenly, without tedious genetic analyses. This occurred when a young boy died during surgery to mend a broken ankle. The child was diagnosed as having malignant hyperthermia, a rare and often fatal response to certain common anesthetics and muscle relaxants. The trait is genetically dominant, which meant that many other family members might have the same problem. The boy's father wrote to hundreds of his relatives warning them. Some discovered that they, too, could be stricken suddenly, and many now wear special bracelets cautioning against treatment with the potentially lethal agents.

DOUBLE-EDGED SWORD

The cases sketched above were all reported in the popular press, and usually the names of the victims were included in the articles to enhance journalistic credibility. Even a large, prestigious medical institute reports names of afflicted families in brochures that it distributes. This type of exposure can make the families vulnerable to employment and insurance discrimination once entrepreneurs begin collecting and selling the information. It is not difficult to understand why an employer might discriminate. By avoiding workers who may have costly health problems, a company can save large sums of money on disability claims and health insurance costs, since those costs are often based on the amount of reimbursements made to employees. The incentive is particularly strong for employers who provide their own insurance coverage. Discrimination based on weight and smoking habits has already occurred, as have genetic examples. In the 1970s a screening program instituted to identify people carrying the sickle cell gene was widely publicized, and employers obtained test results that led to the denial of jobs. More recently, managers of several battery factories determined that lead contamination in their plants posed a danger for the unborn babies of pregnant employees. In a very broad type of genetic discrimination, the companies discontinued employment of *all* women of childbearing age, regardless of whether a woman was pregnant. The companies argued that the rule was for the women's own good, but of course such a decree provided protection from future birth defect lawsuits.

The advantage genetic information gives to insurance companies is also obvious, since they are in the business of playing the odds with people's health. Although DNA testing is not yet in the formula, the stage has already been set: many major companies use factors such as blood cholesterol, blood pressure, weight-to-height ratio, age, occupation, and smoking habits to establish rates. The logic of the insurance companies is clear: those who need and use the insurance should bear the brunt of the cost. For items having unknown costs, the solution is also simple: no coverage. According to 1993 press reports, about 160,000 persons per year are denied health insurance for medical reasons. This number could go up dramatically if genetic screening were included.

Legislators are beginning to recognize the seriousness of the privacy issue. At the end of 1991, expert testimony given before Congress documented improper use of genetic information and emphasized the inadequacy of current laws. But as we wait for protection, large sample banks are being created without much objection. Estimates at the end of 1991 put about 50,000 genetic specimens in academic laboratories and another 50,000 in forensics laboratories. The former number was thought to be growing slowly, the latter rapidly. Recently, the military announced plans to store two million samples for identification purposes, and most states have a mandatory sickle-cell testing program for newborns that places their blood samples in storage. Although each sample bank has a reasonable purpose, the people contributing the samples, as well as their relatives, have lost control over access to their genetic information. No significant penalty is associated with misuse of genetic information, leaving considerable potential for abuse and even profiteering from personal genetic data.

Legal and ethical issues concerning genetic information are being taken seriously by the National Institutes of Health, which is currently funding scholarly research into these areas in conjunction with its Human Genome Project. The Project, along with a parallel program administered by the U.S. Department of Energy, is designed to map all of the genes in human DNA and to determine their nucleotide sequences. Such information will make it much easier to pinpoint DNA defects that cause nearly 4,000 genetic diseases. As a part of this effort, the medical gene hunters hope that citizens will volunteer their DNA and family records to generate data banks that can be tapped for everyone's benefit. However, in the absence of protective legislation, it is unlikely that an informed public will participate extensively in a program that places it at such risk.

———+··

Huntington disease raises several practical considerations. One concerns obtaining good advice, which is important even before a test is done: once you know the future, there is no turning back, and with a disease such as Huntington, the test result can lead to overwhelming depression. To help identify advisors, a list of sources is included in Appendix V. Credentials can be helpful in selecting one, since advanced training in a

DOUBLE-EDGED SWORD

medical genetics program at a major institution is the standard route to proficiency.

In addition to helping you decide about taking a test and helping you interpret the results, a good genetic advisor should anticipate problems with relatives whose blood samples are needed for interpretation of tests. For example, you might prefer to know your future without notifying others in your family, and your relatives may in turn not wish to know their own futures. Then confidentiality would be important. A serious problem could arise if your advisor felt ethically obliged to release the information to a relative who is unknowingly at risk. Even if you refuse to grant permission, the counselor's professional stance might be sufficient to justify such an act. Obviously, you may need a clear, written policy statement from the counselor before having certain tests done.

Another practical issue concerns paternity. Occasionally genetic tests that include samples from child, father, and mother reveal that the legal father is not the biological father. If you have a paternity secret, you should consider the consequences of disclosure before participating in testing.

Still another issue deals with privacy outside the immediate family. Since test results can potentially influence employment opportunities and insurance coverage, the cost of a test can be much higher than the fee charged. But the benefits can also be substantial if test results enable you to treat a disorder or to avoid bearing a child with the disorder. The current privacy procedures used for AIDS virus testing provide a reasonable model: names are not included on samples sent to testing laboratories, so only the physician has the records. Moreover, for AIDS, protective state laws are in place (in New York State at least one privacy lawsuit has arisen from the release of names of HIV-positive persons). Similar protection is unavailable for hereditary diseases, but you can reduce the chance that your name and genetic profile will appear as part of a genetic registry or a research project by having a written agreement that the data not be released.

A major problem arises when we apply for insurance, since health and life insurance companies can ask us to release medical records as a condition for providing insurance or for reimbursing test costs. Some of this information is pooled and made available to a large number of insurance companies by the Medical Information Bureau (MIB) to prevent insurance fraud. According to MIB, health reports in its files are deleted after seven years. Moreover, participating companies are not allowed to decline coverage or charge a higher rate based only on MIB information. However, if MIB or a comparable organization kept information permanently and combined it with family relationships, the insurance industry would have a potent way to increase profits. Some people fear this and opt to pay all of the test costs themselves rather than seek insurance company reimbursement. Whether personal payment will maintain privacy is not clear, since the test result still enters the medical record.

Genetic testing of children deserves a special comment, since they cannot give informed consent. Consequently, serious ethical issues are associated with telling children about future diseases or with using them for genetic research. For a malady such as Huntington disease, testing children is generally considered unwise. Beware of advisors who advocate testing children for diseases lacking cures.

Another consideration concerns test accuracy. Regulations exist that could ensure high quality genetic laboratory testing. However, the National Academy of Sciences reported in 1994 that the regulations were not being enforced. Until they are, consumers who are worried about accuracy must submit the samples to several different laboratories.

Another type of uncertainty is based on the nature of the tests. For example, the accuracy of RFLP-type tests depends on how close the polymorphic DNA marker is to the disease gene and on whether multiple genes are involved in the manifestation of the disease. In some cases, such as the RFLP tests for Huntington disease, the chance of recombination, and thus misdiagnosis, was slight. Instead, the problem was having information from enough generations to know which form of the polymorphic region was tracking with the disease. Now that the defect in the gene has been found, the test can be changed to look directly at the nucleotide sequence, just as described for cystic fibrosis in Chapter 1.

A final issue raised by Huntington disease is the importance of families in promoting research. Nancy Wexler's father played a pivotal role in raising research funds and in getting scientists to work on the disease.

Taking an active role is especially important for the rarer diseases, since federal research funds are becoming increasingly scarce. Many of the volunteer organizations listed in Appendix V act as fund-raisers and advocates, in addition to distributing technical information and providing emotional support.

Now that we know the molecular defect for Huntington disease, cystic fibrosis, and a number of other maladies with seemingly simple inheritance patterns, the next step is to repair the defect. That's a tall order for a true repair of the trillions of cells in a given individual. However, disease often damages one organ much more than others, and that greatly limits the extent of gene therapy needed. The first primitive experiment is described in the next chapter. Within two years this single case had expanded to tests with almost sixty patients. It is only a matter of time before DNA-based methods replace much of the surgery now practiced by physicians.

Practical Considerations

- Check family history for genetic diseases (see Appendix III).

- Before elective genetic testing, consider risks of uncovering an incurable disease.

- Sharing genetic information with family members is appropriate when it helps them avoid risk or seek treatment. However, for many genetic diseases the information may not be appreciated, especially if nothing can be done.

- Keep genetic data private. This includes information about miscarriages, since multiple miscarriages sometimes suggest a genetic problem and confidentiality may not be vigorously maintained by city agencies collecting the information. In particular, avoid allowing miscarriage information to be added to an application for a birth certificate.

- Before agreeing to a genetic test, know who is going to pay for it (for asymptomatic people, a genetic test is not medically necessary for diagnosis or treatment of an illness, and insurance companies may not cover it).

- Before elective genetic testing, consider insurance and employer ramifications as a part of the cost/benefit evaluation.

- Before genetic testing, obtain a written statement from your doctor regarding disclosure. (A 1989 survey revealed that for certain conditions some geneticists feel obligated to disclose information to relatives, insurers, and even employers.)

- Consider conflicts of interest your doctor might have if you are advised to have a genetic test. These include ownership of testing laboratories, patent holdings, test reimbursement to cover counseling costs, and subsequent use of tissue samples, even if anonymous.

- Avoid testing children for diseases having no cure. Placing a stigma on a child can be very disruptive to a family.

- In the future, you may be advised to take a battery of tests. Be careful with such tests because obtaining adequate counseling is difficult (time consuming) when many different diseases are considered simultaneously.

- To check the accuracy of your record at the Medical Information Bureau, write MIB, P.O. Box 105, Essex Station, Boston, MA 02112, for the necessary forms. According to MIB, about 4% of those requesting records find errors.

Fixing Bad Genes

As a baby, Ashi seemed to be sick more than other infants. Her parents sought expert medical advice, and when the definitive diagnosis came back, their despair deepened. ADA deficiency, they were told, a rare blood disorder. There are probably more people studying it than those who have it. Her body lacks the enzyme called ADA (adenosine deaminase), and that stops her from mounting the immune response needed to destroy invading microorganisms. Even the innocuous microbes that we encounter every day will be a problem for Ashi.

In the past, children like Ashi rarely reached their second birthday. Their short lives were a constant battle with bacterial diarrhea and other infections. Today, antibiotics help, but they are not a cure. The only real hope involves manipulating cells and genes. The ADA enzyme normally converts one molecule (deoxyadenosine) to another (deoxyinosine). For most cells, the absence of ADA isn't a serious issue, since backup systems are able to make the second molecule (deoxyinosine). However, without the conversion process, the first molecule, deoxyadenosine, accumulates to abnormally high levels. That raises a special problem for cell types critical to the immune system, T-lymphocytes and, to a lesser extent, B-lymphocytes. These cells contain a particularly active kinase, an enzyme that changes deoxyadenosine into a third molecule, deoxyadenosine triphosphate. Consequently, if deoxyadenosine levels build up, as they do when ADA is missing, then lymphocytes make high levels of deoxyadenosine triphosphate. An overdose of the triphosphate kills the lymphocytes. Without the lymphocytes, a person is immunologically defenseless.

The cure for ADA deficiency is conceptually simple: supply the patient with normal T- and B-lymphocytes. Since the lymphocytes are blood cells that originate in bone marrow, a marrow transplant from a healthy person can sometimes cure the disease. In about one-third of the cases, a transplant does solve the problem. However, tissue incompatibility often prevents marrow transplants from working—the body kills "foreign," transplanted cells when it identifies them as being from another person.

Gene therapy provides a way to bypass the tissue incompatibility problem. The idea is to remove blood or bone marrow cells from an afflicted child, place normal ADA genes into these cells in the laboratory, and then inject the engineered cells back into the patient's body. In principle, gene therapy could cure any ADA-deficient child.

At the time of Ashi's diagnosis, gene therapy hadn't been sanctioned for any disease. However, biologists had been waiting eagerly to test their ideas with an ADA deficiency case. This disease seemed especially suitable for gene therapy because it affected only lymphocytes: only these cells, not every cell in the body, would have to be repaired. Moreover, successful therapy would not require ADA production to be at precisely the normal level, since doctors had already cured children with bone marrow transplants that produced widely differing levels of ADA (having a broad working range provides an important margin of safety because geneticists cannot yet control levels of protein production with gene insertion strategies). Cured cells were also expected to have a growth advantage over sick ones; thus, the cured cells, even if initially present at low concentration, might eventually outnumber the sick ones. Finally, gene hunters had cloned the ADA gene—that is, located and physically isolated it from other human genes; that made the gene ready for insertion into human chromosomes.

French Anderson, Michael Blaese, and the other biologists promoting ADA gene therapy had to convince scientific experts to allow a human experiment. Anderson and Blaese argued that they couldn't cure all ADA-deficient children by bone marrow transplant, since suitable donors often can't be found. Thus gene therapy should be tried.

DOUBLE-EDGED SWORD

Shortly after the ADA protocol began to work its way through committees, other biologists discovered a way to supply ADA directly to lymphocytes through the blood. Direct injection of ADA hadn't worked previously because the ADA protein broke apart too quickly in the bloodstream. Little of it ever got to the lymphocytes. But ADA coated with polyethylene glycol lasted for days, and weekly injections of the coated enzyme gave some afflicted children nearly normal lives. The urgency for ADA gene therapy seemed to disappear. Opponents of the project, who had previously argued that the whole idea was premature, called ADA gene therapy unnecessary.

It was within this context that Ashi, at age two, began receiving the coated enzyme. Her ability to fight disease improved markedly, but near the end of the first year of treatment, her T-lymphocyte population plummeted. Infections followed, and it became clear that something else had to be tried. Experimental gene therapy was approved for Ashi with the stipulation that she continue to receive weekly injections of coated ADA.

The biologists proposed to use a virus to deliver copies of the ADA gene into Ashi's system. Viruses, which are essentially pieces of nucleic acid (DNA or RNA) surrounded by protein shells, enter living cells and reproduce in them. A suitable virus could be made by joining the ADA gene to a viral nucleic acid and coating this recombinant nucleic acid with viral proteins. Then blood cells would be removed from Ashi, infected with the virus in the laboratory, and finally returned to her. To prevent spread of the virus throughout Ashi's body, the biologists would use a crippled virus, one that lacked genes needed for spreading.

A type of virus called a retrovirus was chosen to deliver the ADA gene to Ashi. Viruses of this type, which include the one that causes AIDS, encode their genes in RNA. Shortly after a retrovirus enters a human cell, its genetic information (viral RNA) is copied into a DNA form that then inserts itself into chromosomal DNA of the host cell. In the DNA/film analogy used in previous chapters, viral RNA would be like a short videotape whose information would be copied to form a short segment of film (viral DNA). This film segment would then insert into the middle of a long film (human chromosomal DNA). Once viral information, and the ADA gene carried with it, got into chromosomes, it would remain there for the life of the cell, safe from attack by intracellular enzymes. From this protected position the ADA gene could code for the ADA protein that would save Ashi's lymphocytes from self-destruction.

Although insertion of ADA genes into Ashi's chromosomes provided protection for the genes, it also raised potential problems, since the vehicles carrying ADA genes, the retroviruses, insert their DNA copies at many different spots in chromosomal DNA. Insertion could occur in the middle of one of Ashi's many other genes, splitting it in two and disrupting its ability to function (Figure 5.1a). This type of mutagenesis (gene damage) was considered minor: for statistical reasons, most cells would not receive the ADA gene insertion inside genes important for lymphocyte function, and so there would be little effect. However, insertion of retroviral DNA could also occur next to one of Ashi's genes (Figure 5.1b). This could be a problem, because near one end of retroviral DNA is a region called a promoter that can cause an adjacent host gene to "turn on"—that is, to produce large amounts of messenger RNA and thus large amounts of the protein encoded by the gene. High-level production of certain proteins can cause a cell to grow and divide rapidly. This scenario was worrisome, since cancer in some animals has been associated with placement of retroviral promoters near certain genes.

Neither insertional problem can be evaluated with any accuracy. They are among the current undefined risks of gene therapy. When biologists working with Ashi carried out preliminary tests with monkeys, they saw no obvious ill effects attributable to virus insertion. The theoretical problems with gene therapy seemed remote compared to the known, devastating effects of ADA deficiency. Thus, the plans for experimental treatment of Ashi moved forward.

Blood cells were removed from Ashi's body, distributed into laboratory dishes, and infected with retrovirus particles carrying ADA genes. In late 1990 these engineered cells were injected into her body. In a month she showed improvement. Over the next ten months she received additional injections of engineered cells every month or two, and she felt well enough to attend school with other children. A six-month rest from the injections followed, and then maintenance injections

DOUBLE-EDGED SWORD

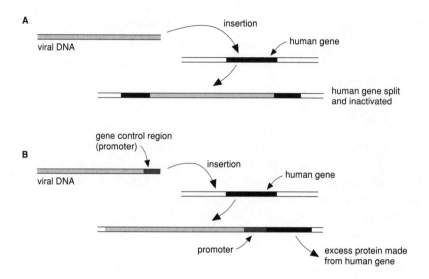

Figure 5.1. Effect of retroviral insertion on host genes.

(a) Splitting a gene. Occasionally a retrovirus inserts its DNA into the middle of a gene in the host chromosome, preventing the production of the normal protein product of the gene.

(b) Increasing expression from a gene. Many retroviruses have, near one end, a region called a promoter that causes large amounts of RNA to be made from adjacent DNA. If viral insertion places a promoter next to a human gene, the virus can cause production of large amounts of the protein normally encoded by the gene. This sometimes leads to uncontrolled cell growth.

resumed at three-to-five-month intervals. After about a year, Ashi showed, for the first time, an immune response in a tetanus skin test. In 1993 her ADA level was about half that of her father, but she was no more prone to infection than her classmates. Her engineered cells now appear to last at least a year, and surprisingly, they tend to help her sick cells last longer. The ADA experiment with Ashi seems to be well on its way to success.

——+·——

Fixing Bad Genes

Unlike infections, genetic diseases cannot be overcome by our natural defenses. However, in a few cases we can control the symptoms through medical intervention, usually by supplying the normal product of the defective gene. We give insulin injections to diabetics, clotting factor treatments to hemophiliacs, and growth hormone supplements to certain dwarfs. Unfortunately, these measures require repeated treatment over a long period of time. In contrast, replacing the defective gene, or at least correcting for its loss by adding a normal one, has the potential of being a permanent solution.

While Ashi's story is far from finished, its effect on the medical community has been striking: by mid-1992, twenty active gene therapy protocols (experimental plans) had been initiated or approved, and by the end of that year the number was forty-one. By mid-1993 it was fifty-eight (at twenty-six different institutions). These experiments were aimed at a variety of diseases, including forms of hemophilia, skin cancer, liver disease, and other cases of ADA deficiency. The liver cases involved abnormally high levels of blood cholesterol that are affected by liver cells. Diseased liver cells were grown in the laboratory, infected with a virus carrying a normal gene copy, and returned to the patient's liver. By early 1994 the engineered cells had clearly taken hold, and the patient's cholesterol level had improved. Before long, we may see gene therapy extended from life-threatening problems to cosmetic ones, although cosmetic therapy is vigorously opposed by current gene therapists. They argue that such actions might be unsafe, since we are still quite ignorant about what most genes do.

Before the technology can be fully used, many technical and ethical problems need to be solved. One critical problem is that only a small number of disease-causing genes have been isolated and cloned in laboratories (by the end of 1993 the number was less than thirty). Cloning a disease-causing gene can be difficult even after its approximate chromosomal location is known. For example, it took ten years of intensive effort to obtain the Huntington disease gene. Diseases arising from the combined effect of several defective genes are even tougher to pinpoint genetically. As a result, most disease-causing genes are not yet available as discrete DNA fragments that can be manipulated and delivered to cells.

DOUBLE-EDGED SWORD

Another problem concerns the delivery of genes. Defective blood cells and skin cells can be removed from the body, infected with an engineered virus, and then returned in working form to the body. However, cells from most other organs, such as the brain, do not function properly when returned to the body. For this reason, we can expect better methods to be developed for handling cells outside the patient's body. We can also expect specialized viruses to be designed for carrying new genes directly to target cells inside the body. An example is emerging from work on cystic fibrosis. A virus called adenovirus, one of the causes of the common cold, infects only respiratory tissue. Therefore it was chosen to deliver a normal copy of the cystic fibrosis gene to lungs (lung damage is often the cause of death in cystic fibrosis). The recombinant virus was tested on rats, and in early 1992 it successfully delivered a working form of the gene to rat lungs. That encouraged human trials. Late in 1992 an experimental protocol was approved for humans, and in April 1993, a twenty-three-year-old man became the first to receive the adenovirus treatment. At this stage, the adenovirus strategy still requires repeated treatment—this virus doesn't insert its DNA into chromosomes where the cystic fibrosis gene would be protected from attack by intracellular enzymes.

Infectious viruses are not the only way to put genes into cells. An alternate strategy involves coating gold beads with DNA and then using pressure to shoot the beads into skin cells. The bead method has already been used as a vaccine that protects mice from flu virus. After viral DNA was shot into skin cells, the cells made viral proteins that stimulated the mouse immune system to protect against infection by the virus. Another strategy is to coat the DNA with fat molecules that allow the DNA to pass easily into cells. Skin cancer has been treated this way with genes that cause the cancer cells to make a tissue incompatibility protein. This protein causes the patient's immune system to perceive the cancer cells as foreign and kill them. In one patient, the DNA treatment led to shrinkage of tumors.

A third technical problem with gene therapy concerns the short life-span of certain cell types. T-lymphocytes, for example, normally live only a few months. In the ADA case, the gene treatment resupplies new, engineered T-lymphocytes as the old ones die: thus, Ashi is not finished receiving injections. A major advance would be the ability to engineer and deliver stem cells, the long-lived cells that divide to form T-lymphocytes. These cells are hard to obtain from older children, but they can be recovered from umbilical cords. In May 1993 an experiment was started with two ADA-deficient babies who were treated with engineered stem cells taken from their own umbilical cords. In a few years we may see a permanent cure for ADA deficiency.

Harmful genes, as opposed to inactive ones, pose a fourth problem: adding a normal copy of the gene to cells will not eliminate the poisoning effect of a harmful one. Such may be the case for Huntington disease. Imagine for a moment that genes are like radio transmitters sending out signals (messenger RNAs). If these messages could be blocked by jamming the transmissions, we wouldn't need to destroy the transmitter itself. Biologists now have ways to interrupt the transmission of messenger RNA from harmful genes by using another type of RNA molecule called a ribozyme. Ribozymes can seek out messenger RNA molecules from specific genes, lock onto them, and then rip through the target messenger RNA. Once the target RNA is cut, it cannot function; its information is not used (translated) to make a harmful protein. Antisense RNAs are similar—they lock onto a messenger RNA and inactivate it without cutting. In late 1992 an antisense gene therapy protocol for fourteen lung cancer patients gained approval from a National Institutes of Health committee of experts, and a year later a ribozyme strategy was approaching approval for an AIDS study.

The major ethical issue associated with gene therapy concerns genetic changes that are transmitted from the patient to his or her offspring, an event that can occur when genetic manipulation affects germ cells—that is, sperm or eggs. The term germ line gene therapy has been applied to this type of procedure to emphasize its importance. It has the potential of gradually changing the human race in a directed way (this is not the case for the type of gene therapy being done with Ashi, where the procedures alter only somatic cells and do not produce inherited changes). Among the philosophical questions arising are whether children have the right to inherit unmanipulated genomes and whether informed consent should apply to patients who are not yet conceived.

As we learn which forms of particular genes give desirable traits, we

DOUBLE-EDGED SWORD

could gradually move into a world where a nation with control over germ line gene therapy would be able to redefine its population to gain advantage over other countries. In the future, we may see worldwide agreements to ban such technology or governmental efforts to implement it as fast as possible. Since we know very little about the possible side effects of gene manipulation, prematurely engaging in such a "genes race" could be dangerous—it may even be unwise to do away with clearly harmful recessive genes, since some may have hidden benefits. For example, beneficial effects have been suggested for the gene involved in sickle-cell anemia (resistance to malaria).

Fortunately, issues concerning a potential genes race will remain in the realm of science fiction for a few more years while the science of gene therapy matures. However, germ line gene manipulation, which is used extensively with mice, is already a refined technology. Fragments of DNA are injected directly into fertilized mouse eggs, where some of the fragments become incorporated into chromosomes. The baby mice that develop contain a new gene in every cell, including sperm cells and egg cells. The engineered mice then mate, and the new genes pass to the baby mice. So far, several genetic diseases of mice, including growth and fertility problems, have been cured this way. Although the methods are still imperfect, the basic technology could be applied to humans. Many biologists worry that such an experiment could accidentally *create* a severely handicapped child, since we are still quite ignorant of precisely what any specific gene might do.

Other technical strategies are also developing rapidly, and now direct gene manipulation is not required to change the genes of a population— embryo selection is sufficient once desirable forms of a gene are known. For this, the initial step of in vitro (test tube) fertilization is carried out in the normal way by mixing sperm and eggs in the laboratory. At an early stage of development, when the embryo has only four or eight cells, one cell is teased away so its DNA can be analyzed. Cells from many embryos can be examined quickly, and an embryo with the desired genetic constitution can then be implanted into the mother's uterus (removal of one cell at an early stage of development does not harm the embryo). There the baby develops normally, and the resulting child will have the desired characteristics. In late 1992 embryo selection

Fixing Bad Genes

was used successfully for a case of cystic fibrosis, producing a healthy baby for parents who were both carriers of the disease. The same thing was done for a case of Tay-Sachs disease about a year later. Now embryos can be split into several portions, each of which could possibly grow into a child. Preliminary human experiments along this line were reported in 1993.

———+·

In a practical sense, we are far from large-scale implementation of genetic change in human populations as long as current procedures are used. Removing cells from the body and then replacing them requires a complex and expensive technology. Nevertheless, the numbers could reach into the thousands per year if an extensive infrastructure of local clinics were set up. Existing in vitro fertilization clinics could provide the nucleus for such genetic services (in 1990 there were more than 150 fertility clinics in the United States, and together they froze more than 23,000 human embryos capable of developing into children). These numbers will grow as the markets expand (evidence for expansion was apparent in mid-1992 when one of the sponsors made a public stock offering to raise funds for blanketing the country with clinics). However, the real impact of gene therapy will come when viral or other gene delivery systems are developed so a simple injection is sufficient to alter genes. Then the instances of gene therapy could reach into the millions, and the main issue will be determining who has access to the technology.

For the present, it is important to stress that correcting somatic cells (cells other than sperm and eggs) does not rid the body of the defective gene: it can still be passed along to the patient's children. Consequently, if you or members of your family participate in gene therapy, it is advisable to keep the information private for the reasons stressed in the previous chapter. It is also important to emphasize that gene therapy is an invasive, experimental procedure; we do not know how it will turn out for any particular case. Repeated injections of large numbers of engineered cells may be dangerous, since insertion of retroviral DNA mutagenizes (changes) chromosomal DNA. Consequently, gene therapy is currently appropriate only for cases having a very poor prognosis. In light of these considerations, it is not surprising that some families with children needing either tissue transplants or gene therapy have conceived additional babies to supply those tissues (siblings frequently have compatible tissues). This raises additional ethical issues, especially when genetic testing is used to identify tissue-incompatible fetuses that are then aborted. Such situations will also stimulate embryo splitting so that an identical twin embryo can be kept frozen until needed.

Ethics and pragmatism also square off when molecular tests are applied to incurable infectious diseases. In the case of AIDS, the early warning can come a decade before symptoms, and that can assist prevention programs. In at least one country, Cuba, the afflicted are segregated from the rest of society. In most of the world, however, the tests raise difficult choices, as illustrated in the next chapter.

Practical Considerations

- Maintain privacy, since somatic cell gene therapy does not rid the gene from the patient.

- Consider potential problems with the gene delivery system, especially if a virus is used.

- Embryo selection can be used to obtain a healthy baby, thereby bypassing the abortion strategy.

- Use of cloned embryos requires considerable forethought. How would you feel if you could observe your twin, twenty years your senior, growing old? What if one of your quintuplet siblings were grown to maturity by your genetic parents, while you and two others were frozen for a decade as embryos before being sold to other families?

Early Detection of Infection

Jane was a nurse in a midwestern community hospital. One night in the late 1980s, paramedics brought an unconscious young man to her emergency room. The doctor on duty quickly ordered an intravenous tube, and Jane stuck a needle into the man's arm. For some reason, adhesive tape wasn't handy, so Jane had to hold the needle in place with her bare fingers. During the few minutes it took to find tape, blood leaked from the man's arm onto her fingers. Jane didn't pay much attention to the trickle of blood, but the doctor did. He had recently relocated from New York City, where he had encountered many AIDS patients. Any free flow of blood made him nervous. "Blood!" he exclaimed forcefully, motioning toward Jane's hand. Just then another nurse arrived with tape and attached the needle. Jane immediately washed her hands. Several days later, the doctor quizzed the patient and discovered that the young man had AIDS. His blood had exposed Jane to HIV, the AIDS virus. When the doctor told Jane, she remembered that she had been gardening earlier in the day that the patient came in. She had scratched her fingers, and that worried her because the little cuts could have made it easier for HIV to infect her. The doctor tried to be reassuring. The chance of being infected was probably small, perhaps comparable to the risk of being stuck by a contaminated needle. For that, the odds of infection were only 1 in 250 to 2,500.

For several weeks Jane walked around in a daze. In calmer moments she struggled with whether to take the blood test for HIV infection.

DOUBLE-EDGED SWORD

Obtaining a positive result meant knowing for certain how her death would come. HIV attacks and kills T-lymphocytes, and as pointed out in Chapter 5, a person lacking these blood cells is immunologically defenseless. Jane had seen weeks of diarrhea turn muscular young people into virtual skeletons.

A positive test result also meant that other people would know. How would her family react to the news? Would they be frightened of her? When she visited, would they shunt her to a motel instead of allowing her to stay at the family home? And would they help when she ran out of money? She certainly couldn't get any more insurance. At best, she could only hope that her present coverage would be maintained. Drastic reductions had been forced on others. And what would happen if she told the hospital administrators? Would they find her another job that didn't involve patient care? Or would they look for reasons to fire her? Could she get the test results secretly? Perhaps she wouldn't have to tell anyone.

After a month, Jane decided to phone the emergency room doctor. The anxiety of not knowing was too much to handle. She asked him to draw her blood for testing, and he complied.

The standard HIV test is based on the finding that infected blood contains antibodies that bind specifically to parts of HIV (antibodies are proteins that help our bodies recognize and fight infectious microorganisms). A technician in the lab mixed some of Jane's blood with proteins purified from HIV to test for the presence of antibodies. Her first blood samples revealed no antibodies directed against HIV, and this gave her hope. However, several months are often required before an infected person's system produces detectable levels of antibodies against HIV. Consequently, Jane had to pass tests for six months or more before she could feel safe.

Jane also needed to decide about treatment with a drug called AZT. The drug blocked viral reproduction in cultured cells, but it had never cured anyone. For some people, AZT did seem to prolong life by a year or two. Should she start a difficult drug treatment program before knowing whether she was infected? She decided to wait, since the odds were on her side.

At the fourth month Jane tested positive for HIV. A light seemed to go out in her; she couldn't even cry. It was all she could do to drive home and crawl into bed. Hunger pulled her out the next day, and she forced herself to go to work. After a week her numbness began to wear off, and within a month she started AZT therapy. It isn't certain that the drug did much good for her. However, the early warning did give her time for things she'd always wanted to do, like tour Europe. But while there, she felt alone and vulnerable without the support of her family. Sadly, the best part of the trip was returning home. Her world had begun to shrink; soon she would be happy to simply walk around her bedroom or finish a meal.

The early warning also gave Jane a chance to focus her time on her family. Her parents were surprisingly understanding and supportive, although she did occasionally run into insensitive relatives. Two years after her diagnosis, Jane began to suffer from infections caused by normally harmless microorganisms. The vomiting, diarrhea, and high fevers eventually wore her out. One evening, with her family at her side, Jane passed away. The same fate now awaits a million other Americans.

———+·

With AIDS, our molecular tools have created a situation in which a fatal, infectious illness can be diagnosed while the afflicted person still feels absolutely normal. Individual responses to a positive test result vary considerably. For some persons, each day takes on heightened meaning. For others, the response can be suicidal depression. The prospect of depression can be so devastating that some people at risk for AIDS refuse to take the test. Instead, they periodically monitor the concentration of T-lymphocytes in their blood as an indicator of disease. If they are infected, this strategy allows them to pass through the early phase of the disease without knowing their fate. Of course, they miss whatever benefit early drug therapy might provide.

For those who have been exposed but not infected, the HIV test provides a welcome relief. The experience of a New York City physician is a good example. When the doctor was accidentally stuck with a needle containing contaminated blood, she became so worried that she immediately stopped eating in the same room as her young daughter—

DOUBLE-EDGED SWORD

even though she knew that AIDS isn't transmitted by casual contact. Only after she tested negative for a year did the doctor begin to relax. Without the test she might have had a ten-year wait before she could feel safe.

Early diagnosis of HIV infection also allows health care workers to be tested as protection for their patients. Doctor testing gained national attention when a young Florida woman was infected by her dentist. Her illness touched off a storm of controversy over patients' rights to know whether doctors and nurses carry the virus. Although not many health care workers are infected, the numbers are significant: according to press reports, by mid-1991 6,500 of them carried HIV. Since only 10 percent of the total cases are thought to be reported, another 50,000 are probably HIV-positive. The medical community currently considers the chance of infection from health care workers to be extremely small—there are no documented cases of transmission from infected doctors to patients other than those associated with the Florida dentist (by mid-1992 more than 15,000 patients of HIV-infected health practitioners had been tested). On the other hand, health care workers are occasionally infected by patients, mostly through needle sticks and scalpel cuts (through late 1992, the number of confirmed cases was about thirty). Thus health care workers have become very cautious, and everyone handling blood or blood products wears rubber gloves for protection. Some states even insist that boxers take an HIV test before fights and that surgeons infected with the virus inform their patients.

Infected doctors and dentists face a difficult decision. To reveal their infection means certain financial ruin at a time when there is little evidence that they place patients at risk. If more stringent surveillance of health care workers were instituted, what would happen to doctors and nurses exposed to the virus but not proven to be infected? Would they be temporarily blacklisted? That could bankrupt them, since even the rumor of infection can ruin a medical practice.

Another ramification of early detection is an increase in account-ability, as was seen when a Tennessee man with AIDS suddenly col-lapsed on a sidewalk. His fiancée begged an off-duty policeman for help. The policeman responded with mouth-to-mouth resuscitation and only later learned that the victim had AIDS. The fiancée was charged with reckless endangerment for not properly informing the policeman. In this case, the legal system had to grapple with the medical question of whether HIV can be spread by oral contact.

Accountability also applies to sex when an HIV-positive status is concealed from a sex partner (AIDS is spread by sexual contact as well as by blood). According to press reports, by the end of 1992 about twenty criminal AIDS convictions had been handed down. Three hundred other cases of this type turned out to be too difficult to prosecute successfully, since it has often been necessary to demonstrate that the accused intended to harm the victim. Now many states are enacting legislation that requires prosecutors to show only that the accused knew and concealed HIV-positive status. Such legislation could have a radical effect on interpersonal interactions.

Early detection of AIDS is also challenging our legal system in other ways. In the mid-to-late 1980s an accused rapist could refuse to be tested for HIV. His victim then suffered for months from the fear of eventually coming down with AIDS. Charges against attackers were sometimes dropped in exchange for the results of an HIV test. Courts now order tests for certain criminals. But when does the accused become a criminal and forfeit his right to medical privacy? Is it at the time of arrest, conviction, or exhaustion of all appeals?

A different impact is being felt in family courts. HIV tests have told thousands of mothers that they are now dying of AIDS, and many of these women are struggling to provide proper custody for their children (soon the number of affected children in New York City will exceed 20,000). Some of the mothers worry that greedy relatives will claim the children simply to get welfare payments. Predeath court hearings now seem to be the best way to assign proper caregivers for the children. As the AIDS epidemic expands, so must the agenda of family courts.

Illicit testing has also become an issue, since knowledge about a sex partner's encounters and HIV status can be a life-or-death matter. Surveys make it clear that we cannot trust what someone else says about his or her sexual history, so we've become vulnerable to commercial schemes offering information about HIV-status. For example, one company offered identity cards that would certify a person as HIV-free following a blood test. But such certification is valid only if the test proves negative six months or so after the last at-risk encounter. How can the certifying company establish when that encounter took place?

DOUBLE-EDGED SWORD

The same problem arises when individuals use black market test kits that examine saliva or urine for HIV. For these reasons, as well as the dangers associated with the lack of privacy, counseling, and quality control, governmental agencies have sought to block unauthorized testing. The conflict may intensify as nucleic acid–based HIV tests become practical. These tests, which detect HIV RNA and DNA, involve a step in which the positive signal is multiplied a hundred thousand times. This makes nucleic acid tests much more sensitive than antibody tests and shortens the time between infection and detection. When these tests are simple and reliable, there will be strong demand for them.

As our understanding of disease deepens, we may find that some of the issues about foreknowledge raised with AIDS also apply to other infectious diseases. Tuberculosis could become an example. This very serious disease is caused by a bacterium that grows inside macrophages, blood cells that participate in killing pathogenic bacteria. Most of us have effective immune systems that in some unknown way suppress the growth of the bacterium. Thus, many of us do not experience disease, even if infected. An initial infection, however, leaves an easily detectable mark on our immune systems. It is thought that as people grow older, their immune systems often weaken and allow growth of tuberculosis bacilli that had been dormant for years. Such reactivation would explain why elderly people living in the same facility tend to have tuberculosis caused by a variety of different strains rather than by a single outbreak strain spreading through the facility. If results of bacterial DNA analyses confirm the idea that geriatric tuberculosis often arises from bacteria acquired years earlier, persons who have tested positive to tuberculosis infection can anticipate problems with the disease late in life, even though they never had symptoms when younger. This would be a positive development, since these people would know to watch for early warning signs. On the other hand, insurance companies would be tempted to deny coverage or charge very high rates in such cases.

———+·———

Persons at risk of HIV infection are generally encouraged to be tested so they can guard against spreading the disease and so they can guard against opportunistic infections caused by normally harmless microbes. Progress against these infections has significantly lengthened the lives of

HIV-positive persons. No drug eradicates HIV. However, AZT may retard the disease a little, and it seems to reduce the transmission of HIV from mother to baby during or shortly after pregnancy.

Early diagnosis also provides a person with the option to redirect his or her life. We saw a dramatic example in the fall of 1991. A routine HIV test required for insurance coverage curtailed the sensational basketball career of Magic Johnson. Johnson chose, at least initially, to focus his energies on AIDS education and prevention. The publicity associated with his situation had an immediate impact: four days after the announcement of Johnson's infection, a national television network finally felt comfortable advertising condom use for disease prevention. Two years later most of the major networks were airing public service condom announcements commissioned by the federal government.

While personal knowledge of infection is useful and must be shared with sex partners, employment and insurance discrimination makes it necessary to protect the information. Anonymous testing centers are valuable for this. As the AIDS crisis worsens and the need for health services mounts, we should watch for increased governmental follow-up of tests through formation of registries that include names and addresses of persons testing positive for HIV. Whether these records can be kept confidential remains to be seen.

For a variety of reasons we must each take responsibility for our own protection from HIV. Uninfected persons can help protect themselves by insisting that new sexual partners have a series of tests before abandoning condom use (remember that the current antibody tests are not sensitive in the early stages of infection; a person should be considered safe only after testing negative for at least six months following each at-risk encounter). As pointed out earlier, DNA tests now under development should reduce the waiting time since viral nucleic acids can be detected earlier than antibodies directed against HIV.

Uninfected persons should also consider banking some of their own blood prior to elective surgery in case a transfusion is needed. The current HIV tests occasionally fail to find viral contamination in donated blood: in early 1992, one out of 225,000 units of blood was contaminated; the average transfusion requires three units, so the chance of receiving contaminated blood is one out of 75,000. This failure frequency may seem small, but it emphasizes that everyone is still vulnerable. One example occurred in mid-1991 when an incident of AIDS in several people was traced to organ transplants. Tissue of the donor, a young man killed in an auto accident, had tested negative twice in 1985. In response to this error, the National Institutes of Health issued a special request for grant applications to develop more sensitive tests for the virus. This amounted to an acknowledgment that current tests, which include analyses involving DNA amplification by the polymerase chain reaction (Appendix II), are still imperfect.

HIV can also sneak in through human error and lapses in security. In 1987, a woman contracted AIDS from a transfusion given following childbirth. The technicians who normally tested the blood supply were apparently off duty at the crucial time. The woman died of AIDS a few years later. In 1989 an Australian surgeon infected four women in his office, apparently by neglecting to properly sterilize his instruments following minor skin surgery. All four had surgery on the same day as a man who subsequently died of AIDS. In this case the doctor was HIV-negative. Even as recently as 1993, a physician in New York City made the error of reusing needles for flu vaccinations. He mistakenly thought that surface treatment with alcohol was sufficient to sterilize the needles. Nineteen people were involved. That year an even bigger scandal emerged in Germany. Two blood supply companies were shut down for inadequate testing, and suddenly millions of Germans found themselves at risk for AIDS. At least six people who received blood from the companies tested positive for HIV.

Part of the problem in Germany may have arisen from blood purchased in Eastern European countries where medical practices sometimes lack rigor. According to 1993 press reports, Eastern European cadavers can slip into the United States as sources for skin and bone transplants. The Food and Drug Administration announced in late 1993 that it intended to require testing and rigorous record keeping of non-blood-vessel-bearing tissue, as it does for other organs. Thus the government is responsive, but surprises are still arising.

Our early warning system has provided us with some obvious strategies to avoid AIDS. But it is also telling us that for our own good we must convince others to take an active role in stopping the spread of HIV. Everyone needs to realize that the virus spreads by heterosexual contact, in addition to homosexual activities and intravenous drug use.

Indeed, heterosexual transmission is the major route of infection in sub-Saharan Africa. There the consequences have been profound. In 1993, for example, a quarter of the pregnant women in Rwanda were thought to be infected. In the United States, teenagers are particularly susceptible to AIDS, since they are sexually active and since they tend to be careless or consider themselves invulnerable to catastrophic events. According to press reports, a fifth of American high school students have had four or more sex partners. Fewer than half of those who are sexually active use condoms, and one in seven contracts a sexually transmitted disease each year. In response, school systems in several major cities began to distribute condoms in late 1991, having decided that the price of sexual experimentation should not be death. By mid-1993 the program had reached as low as the fifth grade in New Haven, Connecticut. Surveys there revealed that a quarter of the sixth graders claimed to be sexually active. But condom programs are too late for some: according to 1993 press reports, a study of 270,000 American teenagers showed that one out of 500 is HIV-positive. One in 500 is also the frequency for college students. There are no safe havens.

In the next chapter we continue to discuss infectious diseases, with emphasis on tracking diseases, assigning blame, and becoming aware of particularly dangerous situations.

Practical Considerations

- Avoid unprotected sex and shared needles. Use care when handling blood.

- Have new sex partners checked for HIV. Remember that the antibody test for HIV may not detect recent infection.

- Consider banking blood prior to elective surgery.

- If you are pregnant and suspect that you may be HIV-positive, get checked for the virus. AZT treatment could reduce the chance of your baby being infected.

- An HIV test, if positive, can alert you to the dangers of infection by microorganisms that are usually harmless. Some of these microbes can be effectively controlled by drugs.

CHAPTER   SEVEN

Who Gave It to Me?

Pat checked into the hospital for a minor neurological disorder. Her surgery went well, but afterward she didn't feel quite right. On the second day after the operation, she suddenly developed a fever of almost 105°F. No obvious source of infection was seen when she was examined, but within twelve hours she went into shock. Then redness began to creep over the palms of her hands and the soles of her feet.

Working on a hunch, the hospital infectious disease specialist shot antibiotics into Pat and gave her powerful medications to increase her blood pressure. In a day or so she felt better, and within two weeks she was released from the hospital. Since antibiotic therapy worked, Pat probably suffered from a bacterial infection (antibiotics generally do not suppress viral infections). Before the antibiotics took effect, Pat's surgical wound had been examined carefully. It was slightly inflamed, but otherwise it showed few abnormal signs; it certainly wasn't highly discolored or gangrenous. However, the disease specialist found a small amount of pus in the wound and took a sample to test for bacteria. He placed the specimen on agar in a Petri dish and incubated it for a day at body temperature. Small, glistening colonies grew on the agar. All were composed of the same type of bacterium: *Staphylococcus aureus*. Staphylococcus causes toxic shock syndrome, and that's apparently what Pat had.

Toxic shock, often associated with tampon use, is caused by a poisonous protein made by staphylococci. The toxin is so potent that only a little bacterial growth is needed to cause the disease. Within forty-eight hours of surgery, the toxin can damage almost every organ in the body. Fortunately, most of us produce antibodies that protect us from its effects. Lab tests showed that Pat lacked those antibodies.

Ordinarily, the story would have ended here—the patient survived. However, her infectious disease specialist was also an epidemiologist. Epidemiologists make their living determining how infectious agents sweep through a population of people, and part of their job is to guard hospitals against outbreaks of infectious bacteria. Staphylococci, for example, can be lethal even to those of us who have protective antibodies against toxic shock if a multidrug-resistant strain gains a foothold in our bloodstream. Outbreaks are particularly devastating in newborn intensive care units: in a large hospital, they can cost $50,000 per day. Overall, the annual toll of staphylococci to hospitals in the United States is about $1 billion, which is why hospitals expend so much effort to control the organisms. The disease specialist could not rest until he knew how Pat had been infected.

On the day after Pat's crisis, the disease expert following her case happened to run into a stomach specialist in the hospital cafeteria. The two doctors chatted about their recent cases over lunch, and soon the disease specialist was describing Pat and her postsurgical toxic shock. It turned out that the stomach doctor had seen a similar case at a neighboring hospital four years earlier. He had been called in because diarrhea complicated the usual symptoms of shock and red skin rash. Since he was an academic type, he had written a paper about the case and had sent the staphylococcus strains to a local scientist studying toxic shock. He was eager to compare notes with the disease specialist, and the two quickly discovered that the surgeon was the same in both cases. Could the surgeon be spreading the microbe to his patients?

After lunch, the disease specialist started sorting through hospital records. The surgeon seemed to be leaving a trail of staphylococcal infections behind him. But how could you really tell? Many people harbor staphylococci on their skin—finding it on the surgeon wouldn't prove much. However, DNA-based methods for distinguishing bacterial strains had become available, and he thought that he might be able to track some of these transmissions.

DOUBLE-EDGED SWORD

The disease specialist contacted the toxic shock scientist, and together they began to study the bacterial strains. The scientist retrieved the vial of bacteria stored from the earlier case of postsurgical toxic shock and set out to compare it with Pat's strain. He grew both old and new strains on agar plates, some of which contained antibiotics. The strains were identical with respect to drug resistance, including the ability to grow in the presence of the antibiotic erythromycin. Erythromycin resistance was important because the gene responsible for it in staphylococci is often carried by a transposon, a short stretch of DNA that can hop from one DNA molecule to another. As a result of previous transposon movements, the erythromycin-resistance gene turns up in different parts of the staphylococcus chromosome in different strains. That meant that strains could be distinguished by the location of this drug-resistance transposon. DNA analysis, carried out by another biologist, then showed that the bacterial isolates from both old and new toxic shock cases had the transposon in the same location, one rarely occupied in staphylo-coccus. Since the toxin gene also tends to occur at different chromoso-mal locations in different strains of staphylococcus, its location served as a second test for relatedness between two strains. The toxin gene was in the same place in both bacterial samples. These two mapping results argued strongly that the two toxic shock strains were closely related, if not identical.

The next step was to see whether the surgeon harbored staphylococ-cus on his body. The disease specialist swabbed the surgeon's nose and touched the swab to agar plates. Staphylococcus grew on the agar—the surgeon had a chronic, but benign infection in his nose. As with most chronic carriers of staphylococcus, his blood contained antibodies that bind to the toxin protein, protecting him from toxic shock syndrome. When the transposon and the toxic shock genes were mapped in the DNA of the surgeon's strain, they were found in the same rare places as in the bacteria isolated from the two patients. It looked as though the surgeon unwittingly transmitted staphylococcus to his patients. Just how was not clear. Perhaps during surgery he inadvertently touched his own face and then the patient, or perhaps the bacteria spread through the air.

The surgeon was very alarmed that he was the source of these infec-tions. He was treated with antibiotics, and the staphylococcal infection was eradicated from his nose. That ended the epidemic. He has now been followed for eight years, and no other incidents have occurred.

———+··

DNA analyses are helping us track infectious agents that kill premature babies, surgical patients, and users of dialysis units. Sometimes patients carry the organisms, as became evident when epidemiologists looked at infections associated with cataract surgery. In one out of a thousand cases the eye undergoing the operation becomes infected with bacteria. The scientists obtained bacterial samples from infected eyes and then grew pure cultures on agar plates. They also grew bacteria from the skin of the surgeons and from the faces of the patients. Then they transferred the bacterial cells from the agar into flasks containing a liquid growth medium to grow large numbers of bacteria for DNA analysis. When they compared the strains, using DNA strategies similar to those in Pat's toxic shock case, none of the eye cultures matched bacteria from the doctors. Instead, all behaved like bacteria found on the faces of the patients themselves. There was even a case in which two patients treated by the same surgeon on the same day got eye infections. In both cases, the organism responsible seemed to come from the patient. Thus, with eye surgery, the likely cause of infection is a breakdown in sterile procedure: bacteria from the eye, or the area around it, slip into the wound. There the bacteria find a fertile environment and multiply rapidly.

Postoperative infections are only one type of problem posed by bacteria and viruses. Two others are tuberculosis and AIDS. DNA-based technologies are being applied to both, but for very different reasons.

Tuberculosis (TB) is an infectious, airborne disease. We can catch it simply by being in the same room as a person carrying an active infection. Although our immune systems protect many of us from developing an active infection, the disease can be deadly if untreated. TB became front-page news in the fall of 1991 when drug-resistant strains were blamed for more than a dozen deaths in prisons. It is thought that AIDS has made prisoners much more susceptible to TB by lowering their immunological defenses (in late 1991, 20 percent of the inmates in New York State prisons were HIV-positive). The result has been a

DOUBLE-EDGED SWORD

900 percent increase in tuberculosis cases in New York prisons over the last decade. In early 1992, about a quarter of the prison inmates were showing a positive skin test for TB, meaning that their immune systems had encountered the bacterium (the national average among all persons is only 4 percent). By 1993 public awareness of TB had increased to the extent that the federal government began investigating air-recirculation systems in commercial airliners as potential contributors to the spread of the disease. This effort was advanced by a report that a flight attendant infected twenty-three crew members and four passengers. At about the same time, some judges were equipping their courtrooms with plastic tents to isolate persons being considered for tuberculosis detention, and nursing homes were being asked for rooms where recalcitrant patients could be kept under guard, perhaps for as long as a year.

The most troubling aspect of TB is the development of drug-resistant strains. Soon after a person is diagnosed as TB-positive, he or she receives antibiotics, sometimes as many as four simultaneously. Normally the drugs kill a substantial portion of the organisms, the patient begins to feel better, and the incentive to continue the months of antibiotic therapy fades. However, during the treatment period, antibiotic-resistant bacteria arise spontaneously and increase their relative abundance. If the patient becomes lax about taking the medicine, the surviving bacteria multiply. Eventually the patient experiences a relapse due to bacterial growth. Antibiotic therapy is resumed, but at this point the drugs used initially do little good because so many of the bacteria are resistant. The patient is then switched to other antibiotics. If the treatment program is not carefully followed, the bacteria become resistant to the second set of antibiotics. The cycle can repeat until all of our antituberculosis drugs are useless (TB bacilli have been found that are resistant to nine different antibiotics). For persons harboring drug-resistant bacteria, TB can be fatal. Moreover, the bacteria spread by those persons are drug resistant; thus, the next person infected cannot be cured by antibiotic therapy either.

Two major factors contribute to the development of drug resistance. One is the large population of homeless persons in our cities. Malnutrition makes them especially susceptible to TB, and often they have difficulty sustaining antibiotic therapy for the twelve or eighteen months needed to control the disease. The other is AIDS. The depleted immune systems of persons with AIDS provide litttle assistance to antibiotic therapy; consequently, several drugs must be used simultaneously. When the drugs fail, the person serves as an incubator for growing multi-drug-resistant strains of the TB bacillus.

It is now crucial to identify infected individuals, medicate them properly, and prevent them from infecting the general population. Unfortunately, the standard tests for TB (chest X-rays and a skin test for the presence of antibodies against TB) are ineffective with persons having poor immune systems, the very persons at greatest risk. Such individuals often have other infections that make chest X-rays unreliable, and the skin test is useless if an immune response can't be generated. Culturing the bacterium from sputum (a mixture of saliva and mucus) still works, but often three to four weeks are required for a colony to grow on agar.

New approaches using DNA tests are being developed to identify TB bacteria from sputum samples without having to grow the bacterium. If a few technical problems can be worked out, we will have tests for TB that take a day rather than a month. Moreover, once specific bacterial nucleotide sequences are known to be associated with resistance to a particular drug, DNA tests might also be able to tell us quickly which antibiotics will do no good. Such information is currently obtained by testing bacteria for growth on agar containing different drugs, a process that can take months.

Thus, for tuberculosis, DNA tests provide hope for rapid diagnosis and determination of drug-resistant properties of the infecting bacteria. This is especially important for AIDS patients since they are killed rapidly by TB. DNA analyses are also being used to track different strains of the tuberculosis bacillus. In one study the unique nucleotide sequence features of specific strains revealed that members of a group of AIDS patients housed together had what seemed to be the same strain. They apparently caught it from each other rather than from separate outside sources. That finding emphasized that tubercular AIDS patients must be quickly identified and isolated.

———+·———

For AIDS, DNA analyses have focused more on assigning blame for transmission. The most notorious case concerned a Florida dentist named David Acer. One of Acer's patients, Kimberly Bergalis, had the early, flu-like symptoms of the disease in 1987, about a month after having teeth extracted. Two years later she tested positive for HIV. Bergalis denied having the normal risk factors for infection, but authorities didn't take her seriously. She continued to speak out, sometimes vehemently, as illustrated in a public letter to Florida health officials:

> When I was diagnosed with AIDS in December of '89, I was only 21 years old. It was the shock of my life and my family's as well . . . I've lived to see my hair fall out, my body lose over 40 pounds, blisters on my sides. I've lived to go through nausea and vomiting, continuing night sweats, chronic fevers of 103 and 104 that don't go away anymore . . . I lived through the fear of whether or not my liver has been completely destroyed by DDI and other drugs. It may very well be. I lived to see white fungus grow all over the inside of my mouth, the back of my throat, my gums, and my lips. I have it on my tongue. Do you know what it looks like? I'd like to tell you. It looks like white fur that gives you atrocious breath . . . Do I blame myself? I sure don't. I never used IV drugs. I never slept with anyone. I never had a blood transfusion. I blame Dr. Acer and every one of you bastards. Anyone who knew Dr. Acer was infected and had full-blown AIDS and stood by not doing a damn thing about it. You're just as guilty as he was . . . Open your eyes. I also think you're all a complete bunch of wimps. Not one person was ever man enough or woman enough to join in my stand. . . .

Subsequent investigation revealed that in 1986 Acer knew he was HIV-positive. By 1989, rumors abounded that he had AIDS, and ill health finally forced him to sell his practice. In early 1990 the federal Centers for Disease Control (CDC) interviewed him as a possible connection to the infection of Bergalis, and at that time Acer agreed to have blood drawn for a nucleotide sequence analysis of his HIV strain. Based on the results of this analysis, the CDC issued a report in the summer of 1990 suggesting that a dentist might have infected one of his patients. To adhere to confidentiality guidelines, no names were mentioned. About a month later, four days before Acer died, he wrote a public letter to his 2,500 or so patients that appeared to support Bergalis's accusations.

Bergalis angrily called for HIV testing of all health care workers. Politicians joined the cause, and near the time of her death, Bergalis dragged her shriveled body to Washington to testify before Congress. The political pressure led to an examination of patients whose doctors and dentists were known to be infected with HIV.

During the CDC investigation, other patients of Acer were screened (over a thousand by 1993). Half a dozen turned up positive for HIV. The virus was cultured from each, and portions of the nucleotide sequence were determined for a particularly variable region of the viral genome. A diagram, much like a family tree, was constructed to relate the sequences of the virus taken from the different patients and the dentist. HIV from Acer and some of his patients, including Bergalis, clearly clustered together on the tree (Figure 7.1). The virus from a patient who was at risk for AIDS for other reasons did not group with that of the dentist or the other patients. That person probably picked up the virus from another source. How Acer infected his patients is still a mystery.

The Acer case, which is considered by many to be a fluke occurrence with respect to doctor-to-patient transmission, represents the beginning of efforts to use molecular biology to ascribe blame for transmission of a disease. So far, most of the HIV cases, including a suit filed by Bergalis, have been settled out of court, although at least one multimillion-dollar judgement was awarded by a jury. Such contested cases will eventually involve DNA testing to identify the source of infection. Current molecular methods, however, do not reveal with absolute certainty that a particular person transmitted a disease to another. Organisms recovered from two people might be highly related, or perhaps even identical, but that would not prove that a particular person was the source: the infection could have come from someone else. To argue against this possibility, we must know how often the specific strain of pathogen occurs in the local community. If it is rare, then the odds for direct transmission from the accused to the victim can be quite high. However, the necessary molecular information is generally unavailable.

——+··——

We are *not* safe from microorganisms. If AIDS has not made that clear, we need only consider staphylococcus infections and tuberculosis (in 1992, almost a quarter of the TB cases in New York City were caused

DOUBLE-EDGED SWORD

Figure 7.1. Clustering of HIV samples isolated from the dentist
and his patients.

Nucleotide sequences from a small region of HIV were determined for
the dentist, his patients with AIDS (indicated by letters), and other local
persons with AIDS. Positions on the tree represent relatedness of the viral
isolates. The analysis is complicated by the fact that each person contains
a variety of viruses. For each patient in this analysis, the two virus forms
having the greatest difference are shown in the diagram. Those from the
dental practice (open circles) cluster together on the tree away from local
controls (solid circles). An average control virus is indicated by an asterisk.
Other local controls even less related to the dentist strain are not shown.
Two dental patients, D and F, reported high-risk activities, and so they
are shown as solid circles. The figure is adapted from C.-Y. Ou et al.
© *Science* Magazine 256: 1165–1171 (1992).

by drug-resistant bacteria; these infected individuals may soon form reservoirs for infection of the general population). Since it is far better to avoid infection than to win a lawsuit over disease transmission, a few prevention tips are discussed before summarizing blame assignment.

Sidestepping disease-causing organisms requires different strategies for different organisms. For HIV, the usual advice is to prevent direct contact with the virus: avoid unprotected sex and used syringe needles, do not handle blood with bare hands, and store some of your own blood prior to elective surgery. For drug-resistant bacteria, minimize stays in hospitals and similar institutions: when many sick people are confined to the same area, the concentration of infectious organisms becomes very high. Persons who know that their defenses are weak should inform hospital staff that they will need special care and protection. This includes anyone on immunosuppressive drugs and anyone who is HIV-positive. It also includes the relatively rare persons who are susceptible to toxic shock syndrome. The latter should be especially careful about procedures that involve packing a wound with absorbent material, as is often done with cosmetic surgery of the nose.

To minimize surgery-related bacterial infection, it is prudent to evaluate postoperative infection frequencies for the available hospitals and surgeons. Since the relevant information may not be public, you may need to get your family physician to request infection-frequency information for the type of surgery to be performed. Data in which various types of surgery are lumped together may not be useful for comparison because some hospitals concentrate on high-risk surgery while others perform safer operations. With luck, you may be able to get data for the surgeons you are considering. If you are simply advised that infection frequency is too low to worry about, you might wonder why the numbers are not available. Aren't the surgeons and hospitals proud of their records?

If you have the luxury to shop for hospitals, you might also ask about the quality of hospital infection control programs. We depend on these programs to identify contaminated medical devices, particularly invasive tubing (in the United States, such contamination causes 100,000 to 200,000 infections per year, mostly of the lung and urinary tract). You can ask about the number of persons in the control unit, the annual budget of the unit, and the number of staff having doctoral degrees from major medical institutions. And once you're in the hospital, it's a good idea to find out how often your invasive device should be changed and cleaned. Then you and your family can monitor the cleaning schedule.

If avoidance tactics fail, there are DNA methods that can help you attribute blame for transmitting an infection. The tests can show that the strain of organism you caught has a nucleotide sequence closely related to that found in the person you suspect. (The sequences need not be identical because some viruses, such as HIV, change as they multiply in a person.) It is difficult, however, to convince a court of law that a particular person infected you because there are likely to be many sources for the organism you caught. The situation is especially difficult with AIDS because HIV mutates rapidly and people end up harboring a swarm of related, but nonidentical, viruses. Consequently, an out-of-court settlement is probably the best option. On the other hand, if you are the accused, a DNA test can show definitively that the organism you carry is different from the one carried by your accuser.

DNA analyses are not limited to bacteria and viruses—they are also applied to people, frequently to connect criminal suspects with tissue samples left at crime scenes. Databases are being assembled that make the chance of false accusation vanishingly small. Unfortunately, application of the method has not always been rigorous, and that has led to some of the controversy addressed in the next chapter.

Practical Considerations

- Minimize visits to areas of institutions having high concentrations of infectious organisms.

- If you are immunosuppressed and you must enter a hospital, find out how to avoid TB patients.

- When you are in a hospital, find out how often catheters and other invasive devices need changing and cleaning so you or a family member can ensure compliance.

- If you test positive for TB, maintain the antibiotic therapy for the full amount of time recommended to avoid generating a drug-resistant strain.

- Prior to elective surgery, you may wish to evaluate the quality of infection control programs at the hospitals you are considering. This may require help from your family doctor.

- If you have not been vaccinated against hepatitis B, you may want to ask prospective surgeons if they have been, since surgeon-to-patient transmission of this viral disease has been reported.

- When DNA-based methods are needed for assigning blame for an infection, it is probably better to settle out of court because the method requires that a great deal of information be available about the distribution of pathogen strains in the community.

†

Beyond Reasonable Doubt

The police found the woman raped and murdered in her own home. During their investigation they discovered that the victim's grandson had visited the rural house several times during the day of the murder, and he became the prime suspect. He tried to shift the blame, but his stories were inconsistent. The young man was scared, perhaps confused. When he came to trial, the prosecutor characterized his stories as outlandish and contradictory. Witnesses placed him at the victim's home during the day of the murder, but they claimed that others—an insurance salesman and a great-great-nephew of the victim—had been around the house too. It was also rumored that a relative of the accused had left town suddenly after the murder. The case seemed circumstantial.

Then came the DNA evidence. A semen sample had been taken from the dead woman's body and shipped to a lab specializing in forensic DNA fingerprinting. For comparison, blood from the men seen in the area that day had also been sent. The jurors would have to learn how DNA fingerprinting works and then evaluate molecular evidence.

The prosecutor brought in an eminent medical geneticist to describe DNA fingerprinting. The geneticist's credentials were read into the court record, including the fact that his resume was twenty-one pages long. As the prosecutor guided the biologist, both were mindful that the jury must accept the forensic DNA methods as being well established and thoroughly endorsed by the scientific community.

The geneticist began by pointing out that scattered through human

Beyond Reasonable Doubt

DNA are polymorphic regions in which very short, adjacent nucleotide sequences repeat over and over (up to thirty times). Such regions are called VNTRs (Variable Number of Tandem Repeats). For a particular region of repeats, that is, a particular VNTR, the number of repeats differs from person to person, but the number is not unique to a particular person. If DNA were cut at specific points on each side of the VNTR, the distance between the cuts in the DNA would also differ from person to person; the length of the DNA fragment generated by such cuts would differ (Figure 8.1). The jury would need to understand how a comparison of DNA fragment lengths could be used to distinguish DNA molecules obtained from different people.

Since a VNTR is a particular type of restriction fragment length polymorphism, the DNA/movie film analogy used in Chapter 4 can be used again to explain the idea. A VNTR is like a commercial inserted

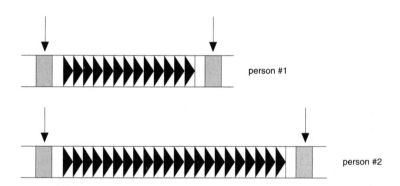

Figure 8.1. Variable numbers of repeated segments create different restriction fragment lengths.

There are regions in human DNA that vary in the number of short, repeated regions. DNA from person 1 has thirteen repeats, that from person 2 twenty-two. Outside these repeats are sequences (shaded) recognized as cutting sites by particular restriction endonucleases. Cutting within those sites (arrows) produces fragments of different lengths for the two people.

DOUBLE-EDGED SWORD

into a television movie, but in this particular type of commercial, the brief message is repeated many times during the advertising break. Local stations use the same brief, repeated message for a given advertising slot, but each station need not broadcast the same number of repeats. Thus the television version of the movie varies slightly from channel to channel as does DNA from person to person.

The witness detailed how DNA is analyzed to relate samples such as semen from a crime scene and blood from suspects. Technicians first remove DNA from the samples and cut it with an enzyme, a restriction endonuclease, to produce fragments of discrete length. For this, an endonuclease is chosen that cuts outside the VNTR. Technicians then compare the lengths of the DNA fragments from different samples by gel electrophoresis. In this method, a slab of material much like gelatin dessert is prepared. The gel slab contains a row of small wells (holes) near one edge, and each DNA sample, which is dissolved in water, is squirted into a different well. An electric current is then run the length of the gel to drive the DNA samples through the walls of the wells and into the gel. The current is turned off before the DNA runs out the other end of the slab, so the DNA fragments are left distributed through-out the gel. Since shorter fragments travel faster than longer ones, the position of each fragment reflects its length. In a sense, electrophoresis is like a horse race that is stopped before any horse reaches the finish line.

Each sample loaded into a well has many DNA molecules in it; consequently, there are many copies of each fragment size. For each fragment size, the copies move together as a band that can be seen if the DNA is stained (see Figure 8.2). The jurors would be looking at bands of DNA fragments to determine which blood sample contained bands that moved the same distance as bands from the semen sample. This match would identify the rapist, who was also likely to be the killer.

When describing how the jurors would see the bands, the geneticist pointed out a special problem: once we treat DNA with a restriction endonuclease, cuts occur at many sites in addition to those flanking a particular VNTR. The many cuts produce hundreds to thousands of different fragments of different sizes; only one or two would be from the particular VNTR we are interested in (in the film analogy, the cutting would occur throughout the film as well as on each side of the

Beyond Reasonable Doubt

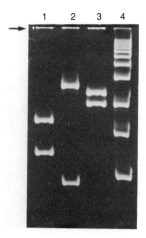

Figure 8.2. DNA bands displayed by gel electrophoresis.

DNA samples were placed in four wells (indicated by arrow) in a thin slab of gel and were then subjected to an electric current. After the current was turned off, the DNA in the gel was stained, and a picture was taken. Lanes 1, 2, and 3 each show two DNA bands; all are of different lengths. Lane 4 contained a mixture of DNA fragments of known size; those bands serve as size markers. Migration of DNA is from top to bottom. Photograph courtesy of Marila Gennaro, Public Health Research Institute.

commercial that interests us). If we simply stain all the DNA in the gel after the electric current has been turned off, we would see so many bands that the DNA would look as if it were smeared through the gel. To distinguish the VNTR band from the many other bands, a sheet of nylon is placed on the gel, and all the DNA bands are driven up and out of the gel onto the nylon. The bands stick tightly to the nylon and maintain the same relative positions as in the gel. Then the nylon sheet is treated with a radioactive DNA, a "probe," that binds only to the DNA in the bands containing the VNTR of interest. That makes the VNTR bands radioactive. The exact position of those radioactive bands on the nylon, and thus in the gel, is then determined by placing the nylon against a piece of X-ray film. The radioactive emissions from the probe bound to the bands of interest expose the film; after development,

DOUBLE-EDGED SWORD

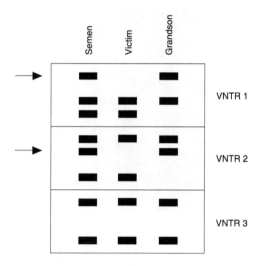

Figure 8.3. Sketch of DNA banding pattern used for DNA fingerprinting.

Dark bands, generated by exposure of X-ray film by radioactive emmisions from DNA probes bound to specific VNTR restriction fragments, differ in position from one person to another. In this case, three different VNTRs were probed. Arrows indicate bands that were uniquely attributed to the rapist, who was presumably the murderer. The figure is a simplified composite to illustrate the text: the relative spacing of the bands is arbitrary.

thin dark bars, about 1/4 inch long and 1/32 inch wide, are seen on the film (Figure 8.3). Since these bars correspond to the DNA bands in the gel, they too are called bands. The jurors would be looking at these dark bands on films.

Another complication is that human chromosomes come in pairs; thus we each have two copies of every VNTR. Sometimes the copies are identical, and then only one band shows up on the film. If the copies differ in repeat number, the probe attaches to two bands, and two appear on the film. Thus the jurors could expect to be seeing simple patterns of bands.

The forensics lab had spent several years testing its collection of DNA probes. Blood samples had been taken from hundreds of people, and company scientists estimated for each probe (and thus each VNTR) how often the different DNA fragment lengths occur among people. From these estimates, odds were obtained for how frequently the DNA from any given person would match the DNA from another person. Consider, for example, a particular VNTR where only 10 percent of the population has DNA with twenty-two repeats (person 2 in Figure 8.1). Then the chance that a random person in the population would also have DNA with twenty-two repeats is 1 in 10 (10 percent); the chance that a DNA sample from a crime scene would coincidentally have twenty-two repeats in this region is also only 1 in 10. Thus, if DNA from a crime scene and DNA from person 2 both had twenty-two repeats in the VNTR being examined, there would be a high probability (9 in 10 or 90 percent) that person 2 was at the scene of the crime.

When several different variable regions (VNTRs) are used in the analysis, the astounding power of DNA fingerprinting becomes obvious. For example, if two DNA samples have fragments that match for two VNTRs (a two-region match) and if the odds for the first VNTR to match were 1 in 10 and the second 1 in 20, then the chance that both VNTRs would match coincidentally is very small, only 1 in 200 (the calculation is made by multiplying the two odds). Likewise, when five regions of DNA (five VNTRs) match, the chance that this happens coincidentally is the product of the five individual matches. That number can be very small, making it extremely likely that the suspect's DNA is the same as that found at the scene of a crime.

In the present case, the commercial lab had three good probes for detecting VNTRs. All three had been described in scientific reports, which the geneticist verified, and the prosecutor entered copies of those reports into the trial record. These papers helped establish that the lab did good work. To further solidify the expertise of the geneticist, as well as to educate the jury, the prosecutor showed a videotape taken in the geneticist's laboratory to demonstrate how each step in the procedure is carried out.

In his cross-examination, the defense attorney first went over the basic technology involved, since some of the jurors might have missed part of the argument the first time around. The essential points were reiterated: (1) human DNA contains short regions, VNTRs, that contain segment repeats; (2) for a given VNTR, the number of repeats differs from person to person, but is not unique; (3) the differences in repeat numbers for a VNTR can be detected by cutting the DNA in specific spots outside the VNTR and then measuring the length of the VNTR fragment by gel electrophoresis; and (4) DNA from a crime scene can be attributed to a particular suspect with high probability if VNTR fragment lengths match with those from the suspect's DNA for several VNTRs. Each step is complex, and therefore each is subject to error.

The attorney then focused on the reliability of the test. He wanted to know whether impurities in the sample could influence the outcome of the test—specifically whether impurities would shift a DNA band into a position that would make one think an innocent person was guilty. The geneticist responded that a person generally knows when band shifting happens, and if it does, one simply disregards the sample. It would be a nonresult rather than a wrong result. Since a person is looking at several probes and several bands, he or she is not likely to be fooled.

Next the defense attorney asked about human error, about the procedures used to keep the samples from getting mixed up. What does the laboratory operations manual tell the technician to do? What safeguards are followed to verify that the right samples got placed in the right order in the gel? The geneticist responded that the loading patterns are listed in the manual, that no deviation is allowed; the samples are always placed in the same order and pattern. A mistake is unlikely. The defense attorney countered by asking how the manual ensures that loading is proper. The geneticist bristled and refused to answer.

The judge called a recess and instructed the geneticist to respond. Apparently the scientist had not read the manual carefully enough to answer this question, and he did not want to jeopardize the technology and the laboratory with a statement that might later turn out to be wrong. That seemed reasonable, so the judge gave him more time to review the manual. The manual did call for the samples to be loaded in a strict order, but no second party was required to watch for human error. That was the main point the defense wanted to make.

Upon redirect examination, the prosecutor also focused on the problem of human error. The geneticist stated that the error rate in his own lab was about one in a thousand. Loading error would not affect a case such as this, because the same mistake would have to be made three or four times to go undetected. The chance of that happening is extremely small, less than one in a billion. Clearly, an observer was not needed to prevent an error from going undetected.

So far, the trial was going well for the prosecutor. His genetics witness was impressive, and that gave his case credibility. His next witness was the assistant manager of the forensics division at the commercial DNA laboratory. The prosecutor entered the credentials of the young man into the record, including the fact that this witness had previously testified at eighteen trials in twelve states. The prosecutor then led the witness through the theory of DNA fingerprinting and into the specific procedures used by his lab. The mundane testimony corroborated that of the medical geneticist and entered more scientific papers into the trial record.

The cross-examination by the defense focused on how the company avoided human error. "That lab manual that you all have," led the attorney, "is not like a checklist that maybe an airline pilot would use where one person stands there and calls off the procedures and some-one—someone else—checks them off as they're done. You don't have that sort of safeguard built into your lab manual, do you, for your procedures in the lab?"

"Well," responded the witness, "I—I don't think that I'd call some-body standing over the shoulder of a scientist a safeguard by any stretch of the imagination."

"Well, excuse me. Do you think that you all are above that sort of safety precaution?"

"Our people in the laboratory are trained, qualified scientists. They're supplied with a protocol and a series of steps that they're to perform in the lab. And as with anybody reading over your shoulder with a news-paper or anything else, it can be quite disconcerting to have somebody standing there over your shoulder. As with any trained professional, it's important that they be allowed to perform their function in a proper manner. . . ."

The defense attorney had made his point: there were no safeguards.

DOUBLE-EDGED SWORD

He then moved from human errors to scientific procedures, particularly those involved in calculating the odds for a match. Here the issue was whether the calculations would be good for a genetically isolated population, as might be the case in the rural community where the murder had taken place. The witness responded that even the Amish of Pennsylvania, a very inbred population, show only slight statistical variation from the overall American Caucasian population. The attorney didn't get very far along this line.

With his redirect examination, the prosecutor tried to reinforce the point that the company was careful about the issue of human error. The assistant manager of the lab stated that three senior scientists at the company go over the data twice for each case before releasing it to the client. Their case load was about sixty to eighty samples per month, enough to provide experience, but few enough so each got handled carefully.

The prosecuting attorney then called in the scientist who actually performed the DNA tests. She pointed out that the semen sample was not a particularly good one: the sperm cells were contaminated by vaginal cells from the victim. She had not been able to remove the contaminating cells, as she sometimes could with a special extraction procedure. Consequently, the crime sample contained DNA from two people. Three tests had been run, one with each of the company's probes (see Figure 8.3). In the first test, the semen sample exhibited three bands following electrophoresis. A pure sample of semen would have produced only one or two (one band is expected if the VNTR is identical on both chromosomes of a pair, two if the VNTR is different). Thus, the sample must have been a mixture from two people, just as expected. Two of the bands aligned with those seen in DNA taken from the victim's blood. The third band had to be from the semen of the killer. The experiment with the second probe showed the same type of result, although the bands ran to different spots in the gel than those seen with the first probe. The third test revealed only two bands from the semen, and they both lined up with the bands from the victim. In that case, the killer's band or bands must overlap those of the victim. Maybe the killer and the victim were related (VNTRs, like any polymorphic region, are inherited).

The scientist had analyzed blood samples from four suspects separately from the evidence DNA. This raised a slight complication, since no two electrophoresis runs are ever exactly the same. However, the scientist had included in each test a series of DNA size markers, which acted like marks on a ruler (see Figure 8.2). By using these markers, she could align bands in different gels. The bands from three of the suspects failed to match those in the semen sample; that is, they failed to run at the same speed during electrophoresis. Those men must be innocent. DNA from the grandson had bands that aligned with those in the semen sample in every case. In the two that had three bands, the grandson's DNA had one band that matched the killer's while the other matched one of the victim's bands. In the third experiment, which revealed only two bands, the bands from victim and grandson matched; no band could be uniquely attributed to the killer.

The prosecutor recalled the assistant manager to apply the statistics and seal the fate of the grandson. The scientist first confirmed the match between the semen sample and the accused. To have a match meant that for the three VNTRs examined, he could identify all the bands seen after electrophoresis of DNA from the semen as coming from a particular suspect or from both suspect and victim.

After explaining the idea of the statistics again, he went to a blackboard and wrote the odds for finding the banding pattern observed. The numbers were 1 in 4,541, 1 in 110, and 1 in 193 for the three tests that they ran. When multiplied together, they equaled 1 in 96 million. This was his estimate of the chance for an accidental match between the DNA from the semen sample and DNA from the accused. He also reiterated that none of the other men seen near the victim's home could have committed the crime: the three VNTRs in their DNA molecules failed to match those of the DNA in the semen sample. No statistics were needed to establish their innocence.

Next the defense called its scientific witness, a young molecular geneticist who was an expert on how contamination affects DNA movement during electrophoresis. His resume was short, and this was his first time before a jury. He was a bit nervous. The defense attorney needed help raising reasonable doubts about the prosecutor's DNA evidence, his 1 in 96 million. The attorney had already established that no second party had verified that the samples were loaded accurately. Now his witness told the jury that a competing commercial lab did use an extra person to verify that samples are loaded into gels properly. Why would they do that if it were not necessary?

The attorney then turned to how the odds had been calculated. The witness reminded the jury that the odds for finding any given fragment length for a particular VNTR are sometimes quite different for different ethnic groups, just as the frequency of genetic disease varies greatly from one group to another. The defense scientist argued that the population used by the forensics lab to get their odds might not be representative of the people living where the murder took place. Moreover, getting the total probability of guilt by multiplying the odds for individual VNTRs would not be valid if there had been inbreeding in this small rural community. If either technicality were true, which nobody knew for sure, the prosecution's 1 in 96 million odds would be a meaningless number. Perhaps DNA fingerprinting was not as iron clad as it seemed, at least in this case.

The banding patterns were next on the defense attorney's list. The jury needed to recall that our chromosomes come in pairs. In some individuals, a particular VNTR is identical in each member of the pair, while in other persons the chromosomal pairs have different numbers of repeats for the VNTR. If the number of repeats is the same, a probe would reveal only one band; if the number is different, the probe would bind to two bands. The grandson's DNA had two bands for each of the three VNTRs tested. The question was whether they were the same as the killer's—that is, the same as those in the semen sample. Since the sample was contaminated by the victim's DNA, there were only two bands to look at, only two bands that could not be confused with those from the victim (arrows, see Figure 8.3). Those two certainly lined up with those from the grandson. But did the killer have only one band for each of those two VNTRs, or did he have two for each, with one coinciding with a band of the victim? If the killer had only one band for any of the VNTRs, then the grandson must be innocent, since he had two. The data were ambiguous at this point.

The prosecutor reiterated that the DNA tests had eliminated all suspects that could be tested except the accused. No fancy statistics were needed to make that point, and no scientific judgment was required. The jury found the grandson guilty. We are left wondering about the relative who allegedly ran off, an issue that was not taken up by the jury.

————+··

Identifying people has long been an important activity in our society. We want to know who the real father is, whether an immigration claim of kinship is true, or whether the accused is really the rapist. For many years, identification efforts have rested on blood-type similarities. However, the blood techniques were never very good because there are only a few different types of blood and because blood samples tend to decompose quickly. Our understanding of DNA now gives us a strong identification method. Unlike blood groups, there are as many distinguishable DNA types as there are people (except identical twins). Moreover, DNA samples last for a long time (the record is 120 million years, obtained by extraction from a beetle embedded in amber), and sufficient DNA can be obtained from a wide variety of body parts in addition to blood samples (a single hair root and nail clippings, to name just two). New techniques will extend the test to the dried saliva left on the back of a postage stamp recovered from a letter. Such versatility and precision have led to a burgeoning industry in DNA fingerprinting: by June 1992 one company had processed almost 80,000 cases at $500 to $1,000 each, and law enforcement groups are now pushing to establish DNA databases to help catch criminals.

One problem is that everyone involved in the judicial system must learn the relevant molecular genetics. Judges need to know what expert testimony to allow, attorneys must draw the right information from the witnesses, and the jurors must make scientific judgments. The jurors have the least preparation, and their task can be very difficult. For example, in one case a prosecution expert had claimed that the odds for a coincidental match between DNA from a crime scene and from the accused were less than one in ten million. A geneticist from Harvard University then argued that the odds were more like one in twenty-four. How does a juror reach a decision in a situation such as this?

The judicial system has formulated several commonsense criteria for jurors to apply. One is that the underlying theory must be scientifically valid, that it is generally accepted by the scientific community. General acceptance is certainly the case for the fundamental principle that each of us (except identical twins) has a unique nucleotide sequence in our DNA molecules and that the fingerprinting method can distinguish the DNA of one person from another. However, experts have disagreed over how to calculate the odds and over specific applications of the method. In late 1991 the controversy exploded. Two prominent population biologists, Richard Lewontin and Daniel Hartl, submitted a paper to the prestigious scientific journal *Science* challenging the methods being used. Prosecuting attorneys were disturbed by the possibility that mere scientific controversy would undermine many of their cases. According to the popular press, the situation became so tense that a prosecutor attempted to privately intimidate Hartl. The Lewontin and Hartl paper was eventually published, and in an unusual move, the editors of *Science* solicited a rebuttal article for the same issue, softening the impact on the judicial system.

A second commonsense criterion is that the method must be reliable: it must not turn up false positives (innocent persons falsely accused). The legal community is sensitive to this issue because forensic tests have not always been foolproof. The paraffin test, for example, turned out to be seriously flawed. This test was widely performed on suspects' hands to determine whether they had recently fired guns. However, contact with tobacco, urine, or fertilizer gave the same result as gunpowder. For DNA analysis, a major problem in reliability stems from the difficulty in recovering adequate amounts of DNA from the scene of a crime. This makes it difficult to remove all of the impurities and still have enough DNA left for the test. The impurities can cause a DNA fragment to move a little faster or slower than it should during electrophoresis. This phenomenon, called band shifting, must be dealt with on a case-by-case basis by the opposing attorneys and their expert witnesses.

A third criterion for jurors is that the method must be properly applied—that is, the odds for a coincidental match must be calculated correctly. Forensics labs have identified specific VNTRs, and they've collected samples from enough people to establish odds. Those odds are valid for the population tested, but some geneticists argue that the odds may not reflect the population at the crime scene. New VNTRs are needed that don't vary from one ethnic group to another. Until this expensive refinement is made, the ethnic problem must also be dealt with on a case-by-case basis.

In addition to the technical problems sketched above, there have been problems arising from prosecutors and forensics laboratories failing to exercise enough care in evaluating evidence. Objections raised by defense attorneys have often been valid. This lack of rigor on the part of prosecutors prompted the National Academy of Sciences to issue a report in early 1992 that emphatically urged scientific oversight and regulation of the forensic labs. Once this is in place, we should see a general acceptance of DNA evidence. Even without laboratory regulation, thirty state supreme courts had upheld the use of this process by early 1994.

—+·

The success of DNA fingerprinting in cases of violent crime has spawned a number of applications. One concerns animal poaching and smuggling. In 1989, the United States Fish and Wildlife Service opened its National Forensics Laboratory in Oregon. Now DNA databases are being established for animals such as bear, elk, and deer. Scientists at the laboratory should be able to identify a specific bear or deer by DNA recovered from bloodstains in a hunter's truck or on his shirt. Blood recovered from a no hunting area could be matched with a steak in a poacher's freezer. Already prosecutors have used DNA fingerprinting to refute a smuggler's claim that a rare South American macaw was the offspring of a captive breeding pair legally in the United States.

Identification of dead persons is another application. A notorious example was confirmation of the death of Joseph Mengele. He was a Nazi doctor who sent 400,000 people to gas chambers in occupied Poland during World War II. After the war, Mengele disappeared into German communities in South America and managed to elude searchers for decades. He eventually died, and a site thought to be his grave was found. DNA was extracted from bones at the grave site, and comparisons between it and DNA from his son and wife indicated that the bones were indeed Mengele's. For most Nazi hunters, that evidence stopped the search, although some still want the bones of Mengele's parents dug up and examined. Another example involved the Czar of Russia. When the Bolshevics took over Russia in 1918, they allegedly killed the Czar and his family. Stories had circulated that a substitution had been made and that the royal family had actually escaped. Recently, the bones were exhumed, DNA was extracted, and DNA analysis was performed. Comparisons made with distant living relatives leave no doubt that the Czar and his family were in the grave.

A much larger identification project is being anticipated by the United States military. During troop induction, a few drops of blood are taken from each recruit, dried onto filter paper, and then added to the recruit's medical record. In addition, epithelial cells are obtained by swabbing the recruits' mouths. These cells are then frozen and stored. At some later date, DNA from these samples can, if necessary, be used to identify casualties by matching it with DNA extracted from body parts found in battle zones.

DNA fingerprinting is also being used to track missing children. The most extensive search has been in Argentina. In the late 1970s, Argentina's military rulers suppressed political dissidents, and more than 9,000 people disappeared. Often babies were spared from the bullets so they could be "adopted" by childless military families. Grandmothers of these children subsequently organized relentless searches for the children. Eventually the lost children began entering school, and administrators noticed five-year-olds registering with questionable birth certificates. School officials notified the grandmothers, who then initiated proceedings to get the children away from their kidnappers. This required scientific evidence, which in the beginning was based on analysis of a blood protein that differs in structure from person to person and from family to family. Matches of blood protein between child and grandmother were adequate for Argentine courts. Now DNA analyses are providing the evidence. So far, matches have been made in about fifty cases. Three times as many are unsolved, a number that is slowly diminishing as the grandmothers track down the children. The children are now teenagers, and they sometimes don't want to hear that "Mom" and "Dad" had their biological parents killed. Thus DNA analyses can be a mixed blessing.

As DNA tests become standardized and as databases grow, nucleotide sequence information is likely to be applied in ways we have not yet imagined. In twenty states, the DNA records of felons are being placed on file, ostensibly to help catch repeat offenders. Rapists are the easiest to identify because they leave semen. In early 1993 the felon file was used successfully in Minnesota to identify a rapist, and earlier in Great Britain the male population of a small region underwent a DNA analysis that led to the capture of a rapist-murderer. We may see the day when everyone has a DNA fingerprint on file. Such a file could be put to many uses in addition to identifying criminals. For example, population geneticists could construct massive family trees, and people could easily track blood relatives. We, as well as insurance companies and employers, could learn whether we have relatives already known to have a genetic disease.

———+·

As the number of criminal cases involving DNA evidence increase, more and more of us will be confronted with DNA fingerprinting. All citizens will be affected by the creation of general-population databases to identify criminals, since we will lose control over our genetic information. Those of us selected as jurors will be concerned with whether DNA tests were conducted properly, and we will struggle to determine which expert witness is correct when their interpretations are contradictory. We will need to remember that experts are fallible, especially when an analytic technique is new. Often the errors are subtle, but occasionally they are obvious. For example, in one rape case the crime lab expert claimed to have adjusted for a band shift by showing how another band, one that all people share, was also shifted by impurities in the sample. That approach seemed fine until the expert inadvertently pointed out that he had found still another band to use for correction—and it gave a different amount of adjustment. The expert insisted that one correction was more correct than the other. Fortunately, the prosecutor realized that he couldn't go to a jury with that kind of testimony.

Attorneys are also fallible, and they may not get the necessary information from the witnesses. In some courts, a juror can ask questions in the form of notes to the judge. For example, a juror in the case history presented at the beginning of this chapter might have wanted to ask about the likelihood of the killer having a VNTR with only one band. Knowing that this is rare might have swayed the juror toward the side of the prosecution.

A juror should also look carefully at the data that is presented. Bands in adjacent lanes in a gel either line up or they don't. Your eyes are as good as those of any expert witness. If there are band shift problems that prevent two corresponding bands from aligning, the identity of the two bands can be shown by a mixing experiment—mixing the two samples together should produce a single band if the two samples are identical, and two bands if they're different.

A few readers will be suspected of committing crimes in which DNA fingerprinting can be part of the evidence. Innocent persons should definitely have the evidence sent to a highly skilled crime lab. A properly run test can establish innocence without any statistical consideration. Such an outcome is common: in almost 40 percent of the cases involving DNA fingerprinting, the molecular methods prove innocence. In one New York case, a defense attorney dug through a laboratory freezer to find the semen-stained cloth that had been entered into evidence six years earlier. A DNA test then showed that the semen was not from the accused, and the man was released. Similar events had occurred in about a dozen cases by early 1993. It now seems appropriate to give all convicted rapists DNA tests if semen samples and victim tissue are still available.

If innocence is not proven by DNA fingerprinting, one can ask whether a relative might actually be the guilty party, since VNTRs are inhertited. According to a 1993 press report, this type of mistaken identity occurred when DNA analysis directed a paternity suit against a seventy-year-old man who had a vasectomy thirty years earlier. His son is probably the man responsible for the child-support payments.

Practical Considerations

- DNA fingerprinting, also called DNA typing, is a statistical method: it does not provide absolute proof that DNA left at a crime scene is identical to that of a suspect. Proper use of the method requires understanding the factors that influence the calculations.

- DNA fingerprinting can prove innocence in a crime situation without any statistical consideration.

- DNA fingerprinting can provide unambiguous family relation-ships.

- Try to ensure that samples contributed to DNA typing data banks are not used for other purposes without prior approval.

Body Enhancement?

Ronnie had always been unusually short. This was a concern for his parents, even when Ronnie was a baby. They were worried about him, and every visit to the pediatrician included the hope that he would rank higher on the height charts. They were on the short side themselves, and they knew how minor things, such as driving a car or watching a movie in a crowded theater, are problems for short people. More importantly, they worried about psychological problems Ronnie might develop if he came to feel small and weak in a highly competitive world.

As Ronnie grew older, his parent's desire to help intensified. The feelings were especially close to the surface when they watched him play with friends who towered above him. And they sensed the hurt every time a stranger remarked about how well coordinated Ronnie was, mistaking him for a younger child. Medical experts fed his parents' fears that a continual barrage of comparisons could damage Ronnie's developing self-image. But what could they do? They told people how hurtful size comments are, but they couldn't train the whole world.

Ronnie's mother had even taken him out of toddler gym class years earlier because she thought that there had been subtle height messages in the class. Like many city kids, Ronnie had attended class once a week to associate with children his age. One exercise had asked him to raise his hands in response to "How big is Ronnie?" His mother felt that this gave value to being big, that importance is dictated by size. She decided that Ronnie was being set up for failure—toddler gym class was not for Ronnie.

126

As Ronnie's mother continued to shield him from situations that called attention to height, the protection itself carried the clear message that something was wrong. And other children reinforced the message, as they sought out Ronnie for the three-year-old boast, "I'm bigger than you are." Ronnie gradually became sullen and withdrawn. His fifth birthday was miserable. For some reason he thought that he would suddenly grow taller with the age change, but turning five failed to bring him magically up to his friends' height.

Ronnie's parents ran out of ideas, and they began to look for help. They knew that people of short stature have support organizations, and a few phone calls put them on mailing lists. They learned about medical aspects of growth from the pamphlets they received. Normal growth is dictated by a number of factors, including several called hormones. Hormones are molecular messengers that travel through the blood from one organ to another and alter the behavior of the target organ. For example, when you are driving your car slightly faster than the speed limit, the sight of a police car can cause your adrenal glands to release adrenalin, which then causes your heart to pump very rapidly. In the case of growth, hormones stimulate long bone extension. A deficiency in either the hormones or in their cellular receptors can stunt growth severely.

One of the important hormones, human growth hormone, is made in the pituitary gland, a pea-sized organ at the base of the brain. The hormone travels to the liver, and from there it is thought to stimulate the growth of bones. A hereditary defect in the gene encoding the hormone causes a form of dwarfism. Where a clear deficiency of growth hormone exists, supplying it through injections can produce remarkable acceleration of growth: pituitary dwarfs sometimes gain up to eighteen inches in final height from hormone treatment.

Ronnie's parents hoped that growth hormone would help him, and they pressed his pediatrician for more information. The pediatrician pointed out that there are more than 100 reasons for short stature. In her opinion, Ronnie was short because his parents were moderately short. His legs and arms were of normal length relative to his trunk, and his head appeared to be normally proportioned. He was what experts call "short but normal." Nobody could guess his final height—perhaps it would be between 4 feet 6 inches and 5 feet. He might even show a

DOUBLE-EDGED SWORD

sudden spurt at puberty. The doctor's advice was to work on Ronnie's self-image and forget about supplying hormones, since they might not help a short-but-normal child.

The pediatrician did acknowledge, however, an outside chance that Ronnie's body was not producing enough growth hormone. She arranged for Ronnie and his parents to meet with an endocrinologist, a hormone specialist. The endocrinologist confirmed that some dwarfs are helped by growth hormone injections, and he agreed that Ronnie was short for his age. The next step was to see whether Ronnie had a deficiency of growth hormone.

Measuring the amount of growth hormone secreted by Ronnie's pituitary gland was not easy. Release of the protein happens in several short pulses throughout the day and night, with most release occurring during sleep. Ronnie checked into the local hospital so blood could be taken for analysis every twenty minutes over a twenty-four-hour period. The doctor also injected insulin into Ronnie to force the pituitary gland to release growth hormone. Neither test revealed a clear case of growth hormone deficiency, and Ronnie was officially pronounced short but normal. The endocrinologist pointed out that this was the usual result, since the condition easily found by the tests, pituitary dwarfism, is rare (there are only about 20,000 such dwarfs in the United States).

Nevertheless, Ronnie was considerably shorter than expected from his parents' height. Moreover, the measurement of growth hormone was notoriously imprecise; thus, there was still the possibility that his body was producing abnormally low levels. If so, maybe something could be done. For dwarfs, growth hormone treatment lasted for many years; that would probably be necessary for Ronnie too. He was already five years old, and they had only until puberty to stimulate his growth. At that point his bones would reach their maximum length, and further hormone treatment would do no good. Ronnie's parents felt a sense of urgency about beginning treatment.

At the time, growth hormone was not easy to get. The sole source was human pituitary glands, which could be obtained only from cadavers. Thirty to fifty cadavers per year were needed to supply each child with an adequate amount of hormone. However, the endocrinologist was able to find enough growth hormone for Ronnie to begin receiving injections.

Shortly after Ronnie started treatment, several other boys taking growth hormone therapy came down with a rare, lethal nerve degeneration called Creutzfeldt-Jakob disease. This is one of a family of diseases caused by a poorly understood group of viruses termed slow viruses. It seemed likely that some of the pituitary glands used as a source for growth hormone had come from persons afflicted with this disease—the virus had apparently survived the extraction treatment used to prepare growth hormone. This revelation quickly stopped the distribution of hormone to all patients.

Fortunately, molecular biologists at a California biotechnology company had devised a gene cloning method to make human growth hormone using bacteria. Their general strategy had been to grind up tissue from pituitary glands and extract messenger RNA molecules. They copied the information from the messenger RNA into DNA molecules, which they joined to other DNA molecules capable of reproducing in bacteria. The resulting recombinant DNA molecules were placed in bacterial cells, and a clone of bacterial cells was found that had picked up the gene encoding growth hormone. The clone, which could be grown to very high numbers, was used to produce copious amounts of the hormone.

By the time the cadaver source for growth hormone was shut off, the form made by bacteria had undergone several years of testing. Soon it was on the market. Never again would cadavers be used to obtain the hormone; never again would growth hormone treatments be contaminated with a human virus. Moreover, the synthetic hormone was available in such abundant amounts that many short-but-normal people, as well as dwarfs, could receive it.

Ronnie resumed his weekly injections, and during the first year he had a little spurt in growth. But then he seemed to drop back to the same growth rate he had shown before treatment. After six years of therapy, he reached the four-foot mark. This was hardly what one would call a striking success for years of hormone treatment. The availability of growth hormone, imprecise tests for hormone production by the body, and an incomplete knowledge of hormone action had encouraged an endocrinologist to prescribe years of injections for Ronnie. There is little indication that Ronnie's final height was affected one way or the other by the hormone. Thus an apparently normal child was pronounced ill and placed on medication simply because he fell at the bottom of the height distribution curve.

———+·

Even before the genetic revolution blossomed, the medical community had made considerable headway toward treating some genetic diseases by supplying patients with missing biological products. Diabetes, hemophilia, and pituitary dwarfism are three notable examples. But obtaining large quantities of pure and compatible material had never been easy. Insulin, for example, was initially obtained from mammals such as pigs. Porcine insulin is similar enough to human insulin to work quite well. However, serious problems occasionally arise with diabetics who become allergic to porcine insulin. For them, recombinant insulin made in bacteria has been a lifesaver.

To treat hemophilia, protein-clotting factors have been obtained from human blood supplies. Hemophiliacs, who number about 16,000 in the United States, lack one of several proteins involved in blood clotting. In severe cases, they bleed spontaneously into joints and internal organs; with the bleeding comes crippling deformities and intense pain. Injections of clotting factors, sometimes as often as three times a week, relieved the bleeding problem and doubled life expectancy. But the outbreak of AIDS hit the hemophiliacs hard, since HIV contaminated clotting-factor preparations. Of those born before 1985, virtually all are HIV-positive. Methods were eventually instituted to produce virus-free clotting factors. Nevertheless, many HIV-positive hemophiliacs who were still alive in the early 1990s were very angry at doctors and public officials for neglecting to tell them that clotting-factor preparations might be contaminated. These HIV-infected men expected help with medical bills and restitution for their infected wives and children. So far, a federal jury has found a supplier of clotting factor negligent, and French courts have sent four officials to prison.

A similar problem arose from the contamination of growth hormone supplies by the Creutzfeldt-Jakob virus. Twenty-five children were infected with this fatal disease. According to press reports, two French doctors came under investigation in 1993 for manslaughter charges. Thus we see two examples of how treatments for disease, clotting-factor and growth hormone, took a step backward through contaminated supplies and perhaps negligence on the part of the medical community. Technology cleared up the problem, but not quickly enough. Clearly, vigilance is required by persons who allow body parts or extracts of body parts to enter their bodies.

The limitless supply of hormones created by genetic engineering can cause a different type of problem—personal experimentation. Hormones are natural regulatory molecules generated by the body in exact amounts. Their ability to perform their function, like manufactured medicines, depends on having the right amount present at the right place and at the right time. Copying this process through delivery of synthetic hormones is difficult. Consequently, the possibility of serious side effects is ever present, as illustrated by a case of erythropoietin misuse.

Erythropoietin is important in regulating the number of red cells in our blood. Persons lacking proper kidney function produce little, if any, erythropoietin, and those individuals tend to be anemic. To treat this form of anemia, a company cloned the erythropoietin gene into bacteria and then recovered the hormone from batches of growing bacteria. This process yielded vast quantities of erythropoietin and made it so readily available that athletes began to dope their blood with it. They had long known that an elevated red blood cell count increases physical performance. Erythropoietin seemed to be the answer to their prayers, increasing aerobic performance by about 10 percent. With recombinant erythropoietin, athletes would no longer have to train at high altitudes to get an edge on the competition. Furthermore, erythropoietin injections would not show up during standard drug screening of urine.

Unfortunately, the competitive edge given by erythropoietin had a dark side: an increase in red blood cell concentration can be dangerous, especially when dehydration occurs, as is normal in sports such as bicycle racing and marathon running. The combination of dehydration and excessive red blood cells causes the blood to thicken, and this thickening predisposes a person to stroke, clotting, and heart failure. From 1987 to the middle of 1991, eighteen competitive cyclists died suddenly from unexpected circulatory problems. It now appears likely that many of them had taken injections of recombinant erythropoietin.

More hormones will become available, and new applications will be found that will make these proteins increasingly tempting medicines. For example, in the course of growth hormone use, it was found that body fat tends to disappear while protein content tends to increase. This observation has stimulated an effort to see if the hormone might be a way to reverse the loss of muscle tone that accompanies aging. Perhaps the next arena of controversy for growth hormone will involve a fountain-of-youth effect.

With a little imagination, we can identify other candidates as rejuvenators. One might be superoxide dismutase, an enzyme that breaks down dangerous oxygen species generated in our bodies. The absence of this enzyme is associated with Lou Gehrig disease, a degenerative neuromuscular malady. Would superoxide dismutase supplementation keep our tissues more fit? Already it has made fruit flies more vigorous and longer-lived. Aging might also be slowed by hormones involved in wound healing. Roughly two dozen growth factors have been implicated in this process, and some are being produced by genetically engineered bacteria to assist healing of bedsores and other chronic conditions found in many diabetics. Perhaps at low doses and in the right combination, these factors will prevent some of the tissue degeneration associated with growing old.

——+··——

The major practical consideration with hormone injection is safety. Hormones are very powerful molecules that often cause a cascade of other events. Boosting the level above normal can have serious side effects, some of which may not be apparent until years later. Even when there is a clear deficiency, it is difficult to administer hormones so the dosage exactly mimics the normal release pattern; consequently, imbalances are bound to occur. With situations such as diabetes, the course of action is clear, since death is the result of not administering the hormone. Body enhancement, however, is often a gamble in which neither the odds for success nor the odds for harm are completely known.

Production of limitless supplies of potent biological molecules by recombinant DNA technologies will create markets for molecules which at most require a syringe to administer. The user, rather than a doctor, will be in control. The "expert" advice will often come from teammates in the locker room and from the friend down the block whose wrinkles or flab miraculously disappeared. Reliance on such advisors makes it especially important to read the medical literature on the subject to learn how to watch for dangerous side effects.

An additional problem concerns children—they are incapable of giving informed consent for "enhancement" of their bodies. As the number of available hormones increases, we will see many more parents placed in the same position as Ronnie's parents, wanting to assist their child's development without knowing exactly how. Often there will be no clear answer, since our understanding of child development is far from complete. Consequently, parents will have the formidable task of learning the appropriate biological details so they can evaluate the costs and benefits. In the next chapter, we will see how this need for biochemical information will apply to many aspects of our lives as genetic screening becomes more widely available.

Practical Considerations

- Hormones are powerful drugs that usually have side effects. Benefits and risks must be weighed carefully. If you do take hormone treatments, know the side effects and watch carefully for them.

- Treatment of children requires careful consideration of the psychological as well as the physiological consequences.

- Injection of blood products or extracts from human bodies can be risky because human beings carry many viral diseases. The problem of viral disease is magnified when the products are concentrated from large amounts of tissue pooled from many donors or when donated materials come from countries having substandard medical practices. Molecules made by recombinant DNA methods using bacteria are not subject to these problems.

- Hormones made in bacteria are not necessarily the same as those made in human beings, even though the gene may be the same—some hormones are modified by enzymes unique to human cells after the hormone is made.

---------------------------- † ----------------------------

Toward Universal Screening

Joyce had heard of malignant hyperthermia in nursing school, but she had never seen a case. Sure, she could recite the statistics: 10 percent mortality with an incidence of one in 15,000 to 50,000 uses of anesthesia. And she remembered what the textbooks said to do about it. Still, she was a bit nervous; this was her first shift in postoperative recovery, and there were a lot of mistakes she could make. Her new boss increased her anxiety with a lecture on keeping alert. Then she was ordered to inventory the MH cart, the place where all the emergency supplies for malignant hyperthermia were kept. Her only reassurance was the faded condition of the emergency procedures poster on the wall. This hospital probably hadn't seen a case of MH in years.

Her first patient was a robust seventeen-year-old named Adam, star of the local football team. His leg had been broken earlier in the day by a crushing tackle. Leg surgery went well, and the surgeon had transferred Adam to Joyce's care. She noted his pulse at 104, double-checked his chart, and then found some morphine. Adam would need some pain relief as the anesthesia wore off. After a few minutes, she rechecked his pulse. It had jumped to 120 and was irregular. Something wasn't right. Blood pressure check. It was too high. Adam's jaw looked funny; his neck was too stiff. When Joyce placed her hand on his neck to feel the tight muscles, she was startled by how hot his skin was. It wasn't the right color either—sort of mottled. Textbook information flashed back to her. Irregular blood pressure, mottled skin, late stages of malignant hyperthermia. Work fast. Her eyes focused on the emergency poster. Call for help—that was the first instruction. Two other nurses rushed over, one with an oxygen tank. Roll Adam over, slip the plastic under, pack him in ice before his temperature shoots up more. Ice-cold saline in the IV. Joyce grabbed the dantrolene from the MH cart. That was the only thing that could save him. Add sterile water to the powder. Shake it. Why won't it dissolve? Another nurse came over to help. Run hot water on the bottle. Somebody called for bicarbonate. Flashback to the textbook: too much calcium released by his cells, all his muscles contract, metabolism goes crazy. Too much heat and too much acid; neutralize it with bicarbonate. Get the dantrolene dissolved and into him; block the release of calcium before cardiac arrest.

Slowly the water in the dantrolene bottle turned yellow as the drug dissolved; the twenty minutes needed seemed like an eternity. But that was soon enough for Adam. The ice had probably helped—sometimes a patient's temperature can jump almost 2 degrees every five minutes.

Joyce had passed the test. Now Adam would spend a couple of days in the intensive care unit, his vital signs would be monitored constantly, and dantrolene would be kept in him for a while to prevent another attack. Two weeks later, the surgeon conferred with Adam and his family. The doctor was certain that Adam had suffered from malignant hyperthermia. Two things caused the episode: the anesthesia given him during the operation and a hereditary predisposition. The calcium channels in his muscles were probably not quite normal, and the anesthesia kept them in the open position. That caused his muscles to contract. Adam should never again have that type of anesthesia.

In half the cases of malignant hyperthermia, hereditary predisposition comes from a dominant gene. That would mean that one of Adam's parents was susceptible to the disease. His brothers and sisters could also be affected, as could his aunts, uncles, and cousins. They should each be checked for the genetic defect, a point the doctor emphasized with a story about an Australian family that lost ten members to malignant hyperthermia. It was only with the eleventh that something was done. Of course, that was thirty years ago. Now most hospitals are better prepared, but still, Adam was lucky to have had an alert nursing staff.

DOUBLE-EDGED SWORD

In the end, many members of Adam's family did get tested. They had muscle tissue removed surgically, and then a laboratory determined whether a common anesthetic, halothane, caused the tissue to contract unusually. Many of those who tested positive now wear a medical warning bracelet to prevent them from inadvertently receiving the wrong anesthetic during an emergency operation.

———+.———

Although the genetic defect responsible for malignant hyperthermia is often a dominant characteristic that can be traced through families, detection is not easy. The symptoms frequently vary, and a person can undergo surgery many times with no adverse effect. Indeed, about half of the cases turn up without a known family history. Even the muscle biopsy is less than ideal: it is expensive, requires anesthesia, and is not performed at all medical centers. A simple DNA test would be especially useful for this trait.

Pigs also suffer from malignant hyperthermia. In certain inbred strains of swine, homozygous individuals (those with two copies of the defective gene) are so susceptible that the disease can be brought on by stress alone. Work with pigs has helped biologists locate the responsible gene, and subsequent gene mapping has led to a human DNA test thought to be 97 percent accurate for some families. Refinement of this test should make widespread DNA screening possible.

Malignant hyperthermia is only one of many acute genetic disorders that can be uncovered by DNA tests. Two others are glucocorticoid-remediable aldosteronism and long QT syndrome. In the former, the front part of a gene that controls the production of a stress hormone (cortisol) fuses to the body of another gene that produces the hormone aldosterone. This genetic rearrangement results in excessive production of aldosterone. That in turn affects salt balance, causing high blood pressure and irreparable damage to blood vessels. When the disease is caught early (before age twenty), it can be effectively treated with drugs that block the action of aldosterone. In the case of long QT syndrome, an otherwise healthy person finds that one chamber of his heart suddenly begins beating at up to 300 times per minute. The first symptom of long QT is frequently sudden death, sometimes brought on by stress as minor as the ringing of an alarm clock. Medication is available to forestall this.

Toward Universal Screening

Less spectacular but more widely useful tests are emerging for common ailments usually associated with aging. These diseases, which include heart disease and cancer, tend to be influenced by complex networks of genes that interact in subtle ways. Sometimes a genetic defect makes a person particularly susceptible to dietary or environmental factors that then lead to disease. One of the better understood examples is coronary artery disease. For many years, physicians wondered why some very fit individuals have heart attacks at an early age, while others, who smoke and overeat, live into their eighties without suffering heart problems. We now know that blood cholesterol levels are often high in heart attack victims. Arteries that supply blood to the heart muscle often choke from cholesterol-containing deposits.

Cholesterol is a fat, which like oil, dissolves poorly in water. This creates a transport problem: our bodies are composed largely of water, so cholesterol cannot be easily moved around in a dissolved form like salts and sugars. The body solves the problem with specific proteins that package cholesterol for transport from one organ to another. One type of protein package is called LDL (low-density lipoprotein). When combined with cholesterol, LDL has been called "bad cholesterol" because high levels correlate with heart attacks.

Members of some families have high levels of LDL cholesterol in their blood, and they tend to suffer from heart attacks. This observation led to a genetic study that identified a gene involved in coronary heart disease. The gene, called FH, for familial hypercholesterolemia, encodes a receptor protein that normally sits on the surface of some cholesterol-producing cells. There it acts as a sensor for LDL cholesterol levels. When cholesterol blood levels are high, many of the receptors bind to LDL cholesterol and signal the cells to stop making cholesterol. If the FH gene is defective, the high-cholesterol signal is not received effectively, and the body produces too much cholesterol. When one copy of the gene malfunctions, blood cholesterol concentration is frequently two to three times above normal. Individuals with this condition often experience a heart attack by age fifty. A defect in both copies of the gene leads to a six-to-eightfold increase in blood cholesterol concentration. In these cases, heart attacks often occur during the teenage years.

Cholesterol levels are sensitive to diet and medication. That allows persons known to carry defective genes involved in cholesterol regula-

DOUBLE-EDGED SWORD

tion to delay artery problems. Even more promising is gene therapy, which has now been performed on a woman whose natural FH genes are both defective. The woman had a heart attack at age sixteen and bypass surgery at twenty-six. At age twenty-eight part of her liver was removed, recombinant FH genes were delivered to extracted liver cells, and the engineered cells were returned to her liver. In 1994, eighteen months after the procedure, the engineered cells seemed to be a permanent part of her liver, and her LDL level dropped by seventeen percent. Thus we are beginning to gain control over some forms of coronary artery disease.

Another type of cholesterol-protein package, HDL (high-density lipoprotein), has been associated with reduced rates of coronary artery disease. HDL, the "good cholesterol," appears to be involved in transporting cholesterol to disposal areas in the liver. Several forms of HDL have been found, and the absence of some, apparently through genetic defects, correlates with increased risk of coronary artery disease. Proper diet, vigorous exercise, and cessation of smoking are currently thought to elevate HDL levels.

Less severe forms of cholesterol-based problems also exist. When they are examined genetically, DNA tests will emerge to give early warning to those at risk. By extending the tests to children, we will be able to catch artery problems before they get too far along. Caution is required, however, because such tests might coincidentally uncover predisposition to other diseases. For example, preliminary work on Alzheimer disease suggests an involvement with apolipoprotein E, one of the proteins involved in cholesterol transport. This protein comes in three forms: E2, E3, and E4. Persons with both copies of the gene in the E4 form have eight times the risk for late-onset Alzheimer disease. Even one copy of E4 seems to double the risk. Since Alzheimer disease has no cure, some individuals may not wish to know the full outcome of apolipoprotein E tests.

Cancer is another type of disease in which genetic predisposition is widespread. When we reach adulthood, most of our cells slow their growth and multiplication, presumably because a set of proteins is made that blocks cell division. At the same time, our body cells gradually accumulate mutations in their DNA molecules. One explanation of cancer maintains that mutations gradually accumulate in the genes responsible for blocking cell division. The mutant cells begin to proliferate; as they accumulate more mutations, they multiply faster and faster. Eventually we see an aggressive growth. We don't know how many genes are involved in shutting down cell growth or how many must be disrupted for a cancer to develop. However, the situation is certain to be complex, with some of the controlling genes being specific to particular tissues and others being common to all body parts.

The rare eye cancer called retinoblastoma has provided insight into genetic factors important to cancer. Retinoblastoma develops in childhood, and if not treated early, it spreads from the eye to the brain. Often this hereditary disease passes from parent to child, but sometimes it skips a generation. And sometimes there is no obvious history of disease in the family. Microscopic analyses revealed that in many cases chromosome number 13 is shorter than normal in cancer cells. Since chromosomal shortening is most easily explained by a loss of a section of DNA in the chromosome, the idea emerged that retinoblastoma might arise from the loss of one or more genes that normally block cell proliferation.

Afflicted children generally inherit a defective copy of the "anticancer" gene from one parent and a normal copy from the other. In the malignant cancer cells, both copies of the gene are damaged. These observations led to the concept that predisposition to cancer arises by inactivation of one copy of the gene and that cancer itself develops when both copies are defective. When the second copy of the gene remains intact, the disease skips a generation. In sporadic cases of retinoblastoma, neither predisposition nor the disease is inherited, and the age of onset is generally later than for cancers having a heritable predisposition. Sporadic cancer is thought to arise when lesions spontaneously accumulate in both copies of the "anticancer" gene.

From retinoblastoma came the general idea that cancers arise from defects in several tumor prevention or suppressor genes. Over time, defects in five or six of these genes gradually accumulate in the cells of a tissue and progressively release the cells from normal growth control. This leads first to a benign growth and later to a highly invasive cancer. One of the key anticancer genes makes a protein called p53, which appears to participate in blocking cell division when DNA becomes damaged. Thus defects in p53 allow damaged cells to proliferate, but the defective p53 does not by itself cause cancer. So far, defective p53 has been associated with fifty forms of cancer, including a high percentage of colon, lung, breast, urinary, cervical, and bladder cancer. A form of skin cancer has even been associated with specific nucleotide changes in the p53 gene. Other anticancer genes, MSH2 and MLH1, appear to be involved in repairing DNA damage. When either of these genes is defective, mutations accumulate in many other genes and generate a form of colon cancer. The vast majority of colon cancer is thought to be associated with a defect in either MSH2 or MLH1. Roughly 1 out of every 200 persons carries defects in the MSH2 gene.

Once specific defects of particular genes are associated with specific cancers, then nucleotide sequence analysis of DNA can be used to assess how cancer prone a particular tissue is. Such tests will greatly increase the accuracy of diagnosis of any suspicious growth. For a case such as bladder cancer, cells that are sloughed off into the urine can be tested for defects in p53 and other genes associated with this type of cancer. That will make early detection of bladder cancer straightforward. The same principle should apply to any organ that releases easily collected cells, such as the esophagus and the uterus. Thus, our early warning system for cancer detection is about to take a major step forward.

DNA tests can also indicate whether a particular person is cancer prone. For example, women in certain families are highly susceptible to breast cancer (their chance of being diagnosed by age fifty is 60 percent while for the general population it is only 1 percent). Chromosome analysis of tumor cells indicates that a suppressor gene resides on chromosome 17, and in some cases, genetic analysis can tell a particular woman whether she is at risk. Comparable statements can be made about a variety of tumors and p53, MSH2, or MLH1 status. Those at risk are advised to be particularly alert for abnormal growths.

———+··———

The age of the gene hunter is in full bloom, and reports occur almost monthly about the identification of genes that will lead to diagnostic tests for predisposition to genetic disease. To the cancer, heart attack, and malignant hyperthermia problems, we can add migraine headaches,

Alzheimer disease, and osteoporosis, a degeneration of the bones that afflicts twenty million Americans. Few commercial tests are available, but it is only a matter of time before each of us will be tempted to undergo genetic screening for one or more specific diseases. In general, we will be happy with the results. If the test uncovers a predisposition to disease, we will be glad to learn this in time to do something. If we have no predisposition, we'll be relieved. Thus genetic counselors will have many satisfied customers.

The question we must each ask is whether the odds for having a hereditary predisposition to a particular disease are high enough to warrant the expense of a test. The answer may often lie in our family histories. With malignant hyperthermia, for example, we would ask whether any of our relatives died during or after surgery, particularly from postoperative fever or heatstroke. If test results are positive, the solution is to avoid particular anesthetics. For cancer, we would ask whether any of our relatives died at an early age of particular types of cancer. Positive test results would suggest a need for frequent checkups, aggressive treatment of tumors, and perhaps even preventive surgery. Each disease will require its own specific strategy.

A negative aspect of screening concerns disclosure of genetic information. Until protective legislation is enacted, the practical question is how to obtain the information without its also becoming available to others. At present there is no simple answer. Whether other negative aspects emerge depends on how we as a society respond to the new technology. This is the focus of the last chapter.

Practical Considerations

- Preventive action can be very effective for some diseases and cancers if detected early; DNA tests will allow early detection.

- Examine your family history for specific types of early death to determine whether there are any obvious problems you can address (see Appendix III).

- If you find a history of disease, gather as much information as possible regarding the diagnosis. Check with a specialist, and get your local librarian to help you search the medical literature for the availability of a DNA test for the particular disease.

- Keep genetic records confidential to protect against employer and insurance discrimination—current treatments do not change inherited genes and may not totally remove all manifestation of disease.

Society and the Revolution

In this final chapter we turn to genetic issues that individuals may have little control over. One of the major concerns is the monetary value of genetic information. As a commodity, DNA technology is important for identifying organisms (including people), for predicting future health, and for constructing specialized organisms. It has great potential both for profit and for holding down health care costs, sometimes to the detriment of a portion of the population. As governments try to balance safety, economic interests, and individual rights, they will find that the correct course of action is not always obvious. New ethical problems will arise, and age-old racial competition will cloud scientific and economic decisions. Thus an informed public is needed to keep society from falling too far behind the scientific advances. Three general areas are discussed as examples of the concerns we can expect as we move into exciting new ways of producing food, controlling disease, and structuring our societies.

One issue currently being debated is genetic engineering of our food. In this area, the tomato has led the way. Since vine-ripened tomatoes contain more sugar than tomatoes that are picked green, they taste better. But ripe tomatoes are difficult to ship, and they tend to spoil quickly on supermarket shelves. Consequently, growers generally pick tomatoes green and then redden them by treatment with ethylene gas. To help maintain firmness, tomatoes are commonly refrigerated, a process that connoisseurs claim eliminates the "backyard flavor." The net result has been unhappy consumers (tomatoes rank lowest among produce items in customer satisfaction).

A solution to the tomato problem seemed to be a tomato that stays firm as it ripens. After conventional breeding efforts failed, a recombinant gene company named Calgene stepped into the arena. Calgene scientists focused their attention on pectin, a large molecule that gives rigidity to tomato cell walls. As tomatoes ripen, a tomato enzyme breaks pectin apart. Although this enzyme action is just right for tomato reproduction, it occurs much too early for mass marketing of ripe tomatoes. Calgene designed a gene strategy to slow the normal breakdown of pectin. The resulting tomato, the Flavr Savr, remains firm for seven to ten days longer than conventional tomatoes. This means that the Flavr Savr can be harvested later; therefore, it should have a better flavor when sold in supermarkets.

Getting the Flavr Savr to consumers was not straightforward, in part because opponents argued that antibiotic resistance problems might arise from eating the tomato. The Calgene genetic construction involved insertion of a new gene into tomato chromosomes. To know that tomato cells picked up the new gene during the construction process, the plant scientists coupled it to a second gene, one that makes cells resist destruction by the antibacterial drug kanamycin. Use of this proceedure, which is standard for genetic constructions with cells of higher organisms, means that Flavr Savr tomatoes contain the kanamycin-resistance gene. When we eat the tomato, we ingest that gene. The *possibility* exists that the kanamycin-resistance gene might not be destroyed by the acid in our stomachs and that it might be incorporated into the DNA of bacteria that normally live in our digestive tracts. That could make those bacteria resistant to kanamycin. In turn, this resistance would make it impractical to use kanamycin to rid the bowel of bacteria, a procedure sometimes used prior to certain types of surgery.

We do not know whether the bacterial drug resistance scenario will ever occur, since it involves many improbable steps. However, if a very large number of people ate Flavr Savrs, even a rare event might show up. The U.S. Government has approached the issue cautiously by designating these genetic changes as food additives. Consequently, the foods must pass safety tests and must be adequately labeled.

By mid-1993 about thirty foods were waiting in line behind the tomato. Some cases involved cross-species engineering, and that created another set of potential problems. One type of objection will come from persons who normally avoid eating certain types of foods for religious reasons. Consider, for example, the problem some groups might have if they suddenly found that the fruit they were eating contained a pig protein. It might not matter that a single gene and its protein product are far from being a pig. Another problem might arise from allergic responses. Some individuals are allergic to certain seafoods, and it is possible that a shellfish protein expressed in a fruit or even in another animal might cause a harmful allergic reaction. Proper labeling and education are expected to minimize problems such as these.

Plant engineering has also raised containment issues. Since the late 1980s, the federal government has regulated the outdoor testing of genetically engineered crops to avoid escape as superweeds. By late 1993, almost 500 had been studied without incident, but some scientists still argue that the weed scenario has not been properly tested. Similar concerns are voiced about the widespread use of plant viruses to engineer plants. The worry is that virulent forms of viruses will emerge and that they will devastate crops. We have no definitive answer for either concern.

Food questions can get intricate, as illustrated by the 1993 approval for use of a bacterially produced cow hormone to increase milk production. The hormone treatment itself appears to be safe, but some people prefer to take no chances. Accommodating them will require costly new record keeping and policing to ensure truth in advertising for old-fashioned milk. Opponents of hormone use argue that it might weaken cows and cause them to require more frequent antibiotic treatment. We certainly do not want to increase our intake of antibiotics, because that can lead to the development of drug-resistant bacteria. Thus we also have to ask whether our testing procedures for antibiotic contamination of milk are adequate. Still another part of the issue is whether the increased milk production will reduce the number of cows and farmers needed to meet the dairy demands of the country. Small dairy farmers worry that the hormone will put them out of business, and so they have been trying to influence protective legislation. At the same time, the producers of the hormone, who have millions of dollars at stake, are trying to eliminate labeling requirements. In early 1994, Maine and Vermont passed legislation requiring labeling of all dairy products from cows treated with recombinant hormone. Consumers were thought to have the right to know what they eat.

————+··

Tampering with immune systems is another issue to track. For many years we have primed our immune systems with vaccines to protect us from infectious diseases. We are now quite accustomed to this type of manipulation, and that makes it easy to accept new applications of immunology. One involves more extensive use of the cowpox (vaccinia) virus. This virus, which helped us eradicate smallpox, produces a rather benign infection but a strong immune response. That makes an engineered cowpox virus well suited to stimulate immunity against preselected proteins. One emerging idea is to use this virus to deliver an infectious, species-specific contraceptive. The concept arises from the observation that sperm proteins sometimes cause an immune response in women, and the resulting antibody action kills the sperm before they can fertilize eggs. Indeed, this is the explanation for some forms of human infertility. Efforts are underway to identify genes encoding sperm proteins. By placing them in cowpox virus and then infecting a woman, an immune response might be elicited against sperm proteins that would, in principle, render the woman infertile. To date, immunity (infertility) caused by sperm proteins delivered by other means seems to last for less than a year. The expectation is that the cowpox technology will create temporary, but relatively long-lasting, contraceptives.

In addition to technical problems associated with unwanted spread of an infectious virus, one can envision social concerns stemming from ethnic differences, especially if the virus were engineered to produce an immune response only against sperm of specific ethnic groups. Simply put, antifertility devices are contraceptives when controlled and administered by the ethnic group using them, but they are genocidal agents when applied from the outside.

Field tests using wild animals are now in progress with the cowpox virus. The best-known experiment focuses on a rabies vaccine rather than on a contraceptive, but rabies is worth watching because the experience gained will have general application. Rabies is a fatal viral disease affecting the nervous systems of warm-blooded animals. Standard vaccines prevent rabies in domestic animals, but they are difficult to deliver to wild animals. To bypass the problems of delivery and expense, virologists designed a recombinant cowpox virus that would express a rabies virus protein. Infection with this cowpox virus leads to immunological protection against rabies. Foxes in Europe became the targets of these antirabies cowpox viruses in the late 1980s. Cowpox-laced baits were placed where foxes would eat them and become infected. According to press reports, the producer of the vaccine expects rabies to be eradicated from foxes in parts of Europe by the mid to late 1990s.

In the United States, the rabies problem has centered on raccoons along the East Coast. The outbreak began in 1979 in West Virginia, and by 1984 raccoon cases were being picked up in southern Pennsylvania, northern Virginia, and Maryland. By 1989 most of Pennsylvania reported rabid raccoons; the epidemic reached the middle of New York and far into New England in 1992. The impact on human beings is just beginning. In 1989 fewer than 100 New York residents were treated for exposure to rabies. The number had grown to over 1,000 in 1992. The extent of the epidemic was driven home in the fall of 1993, when rabies suddenly struck down a farm pony that children had been encouraged to pet while their parents shopped for pumpkins. Saliva from the pony could have infected dozens of children.

For a variety of reasons, American health officials have proceeded more slowly with the cowpox virus than their European counterparts. One worry is that the vaccine may be too good. If rabies is one of nature's ways to control the raccoon population, eliminating the disease could give us a raccoon overpopulation problem to deal with next. Nevertheless, if the epidemic continues to expand, it is likely that American field tests will be conducted. These tests will give us a better understanding of cowpox dynamics for many immunological applications.

—+—

On a different front we have the Human Genome Project, a massive, multinational effort to determine the nucleotide sequence of human DNA molecules and to locate all human genes. Having such information

DOUBLE-EDGED SWORD

is equivalent to possessing coded blueprints for constructing a human being. Even the very fragmentary information we have already obtained has allowed us to define many nucleotide-sequence alterations that lead to genetic disease. It has also enabled us to identify genetic risk factors for a few complex diseases such as Alzheimer disease and several types of cancer. Having the complete set of gene information will make molecular genetics the basis for prevention and treatment of many, many ailments. But with these changes in medicine will come new social dilemmas.

The immediate problems of prenatal (fetal) genetic screening, especially when coupled with abortion, revolve around ethnic diversity: genetic diseases are distributed unevenly among groups of people. For example, cystic fibrosis is prevalent among Scots and English, PKU among Irish and Polish, Tay-Sachs among Ashkenazi Jews, and sickle-cell anemia among Africans. Thus different screening programs are appropriate for different groups of people. Determining whether a particular screening program is worth the effort and expense will be a value judgment made by the controlling groups. Having the program administered by the target group will greatly increase its chance of success. We have already seen that a Tay-Sachs program developed in Jewish communities for Jewish communities is very successful; in contrast, the imposition of sickle-cell testing on African-Americans in the 1970s led to abuse, mistrust, and accusations of genocide. Thus ethnic strife within societies will have a strong influence on screening programs and vice versa.

As screening programs become more sophisticated, the definition of birth defect will broaden: couples will want to use screening/abortion to strengthen their families both physically and mentally. What we now call an average child may eventually be considered defective. At the same time, other advances in molecular biology will improve the length and quality of life for those born with genetic disease. With cystic fibrosis, for example, new antibiotics have reduced the severity of lung infections, and many afflicted individuals now live nearly normal lives into their thirties. Babies born today with cystic fibrosis may live into their fifties. Comparable statements can be made about hemophiliacs, PKU babies, and sickle-cell infants. Governments will have a problem: should they hold down health care expenses by using screening/abortion? The cost will be loss of potentially valuable members, persons such as Abraham Lincoln, Lou Gehrig, and the cosmologist Stephen Hawking.

Even if abortion is discouraged, diagnostic screening will eventually be coupled to in vitro (test tube) fertilization. "Superior" embryos will be identified and implanted into mothers to develop. Here the important technology to watch will be new methods for generating large numbers of eggs from a woman, since the chance of finding desired embryos increases as more are examined.

A different set of issues arises from genetic screening of persons already born. Companies are selling instruments that rapidly determine nucleotide sequences, and soon technical innovations will allow even small clinical labs to obtain sequences for clients. In late 1993, a small biotechnology company in Maryland announced plans to offer DNA testing for cancer risk. Their intent is to sell interactive software to physicians and hospitals for collecting family medical histories. Company personnel will then assist in determining whether DNA follow-ups would be useful for the patient and family members. Confidentiality matters would be left to the physician and the patient, much in the way they are currently handled for AIDS (decentralizing the coupling of names and genetic data is important for safeguarding confidentiality). We must ask whether that is enough. Will society protect its families from being branded as "diseased" and subjected to financial discrimination? Will governments enforce long, mandatory prison sentences for improper release of genetic information by health care workers or theft of it through computer break-ins?

The situation will become even more complex as segments of society begin to require genetic disclosure. Life insurance companies already consider certain genetic factors when issuing policies, and they will certainly want to take advantage of new information. Perhaps insurers will ask a person to produce genetic data to prove that he or she is a good risk. Genetic disclosure may also be required for certain occupations. For example, the public may demand evidence that candidates for high office have no genetic predisposition for mental illness; potential airline pilots or bus drivers may have to show that they are not predisposed to certain acute, hidden heart ailments.

DOUBLE-EDGED SWORD

Even courtship could become a battleground for disclosure: predisposition to a serious illness of midlife could cool the flames of romance. The prospect of handicapped offspring may eventually block potential matings, either from choice or from societal pressures. The latter is already happening among a group of Orthodox Jews who decided to eliminate recessive disorders such as cystic fibrosis, Tay-Sachs, Canavan, and Gaucher diseases from their population. Leaders visit high schools having high concentrations of Orthodox students and tell them about their DNA screening program. After tests have been performed, couples planning to date can learn if they are genetically incompatable. In 1992 about 8,000 young people were tested. We must all watch this experiment closely, since it could be a window on the future. In particular, we need to see how well confidentiality is maintained, how a society deals with diseases that sometimes have only minor symptoms, and what happens to "genetic wallflowers," persons who cannot find spouses because of genetic branding.

Some discrimination problems will be solved by adding, replacing, inactivating, or stimulating specific genes. We are already seeing engineered viruses carry specific genes into bodies to repair malfunctioning organs. Gene therapy will also allow individuals to optimize health and performance where no obvious defect exists. Although the technology is still quite primitive, it will eventually involve little more than removing a vial from a refrigerator, taking a drop of virus-containing liquid from it, and then placing the drop in the patient's mouth. At that point cost will be the main factor restricting access to genetic manipulations.

Cost may be more important than we first realize. As long as sophisticated laboratory skill is involved, cost will be so high that only the elite will have access to genetic medicine. Preferential access to high-quality medical care has long been common, but with genetics the issues are slightly different: access to germ line gene therapy, for example, could permanently alter a family's characteristics. Money could produce a master race. Will governments have the resolve and cash reserves to ensure equal access to all, or will legislation and stiff penalties forbid changes in germ cells?

Perhaps it is a bit extreme to warn against nations becoming involved in the genetic equivalent of an "arms race," but leaders have long felt the need to restructure populations. Hitler's effort to develop a master race was but a recent example, and there is little evidence that modern efforts will be any more sensitive to human rights than his (ethnic cleansing is still a popular concept in the Balkans). It would be wise to consider controls before we have the technology to create superbabies.

With nuclear weapons, the United States and the Soviet Union took control by sheer force. A power strategy may not work well with genetics, since DNA technology is easily decentralized and since its slow evolution lacks the political impact of a nuclear explosion. Moreover, sophisticated knowledge by the public is needed to fully grasp the implications of genetic issues (everyone knew the immediate consequences of a bomb). Consequently, effective control over genetics is more likely to come from superior knowledge. Although the United States now dominates molecular genetics, maintaining that position requires more than simply funding the Human Genome Project. It also requires maintaining a scientific infrastructure that can develop new directions and use the information generated. If this is not maintained, other groups will be the benefactors of the information. Politicians need to understand that new directions come from people such as Max Delbrück, not just from nucleotide sequencers. Today's tight money and focus on applied research makes the type of basic work carried out by Delbrück difficult to fund, and that in turn makes basic research a tough business for individual scientists. For survival, some of our research institutions are making deals with foreign interests for research support in exchange for commercial rights over future discoveries. If we want to guide the genetic revolution, we must make basic research easier to conduct.

Molecular genetics will change our lives. For cancer diagnosis alone, we will all be thankful, and we will want more research and even better solutions to our genetic problems. But we cannot give free rein to those who would apply molecular biology for profit—too many families can suffer irreparable economic damage. With this book I have tried to provide a base for individual action. As a society, we need to focus on doing what is right, not what is expedient.

---------------------------------- † ----------------------------------

Additional Reading

Books That Introduce Specific Genetic Diseases

Beaudet, A., C. Scriver, W. Sly, D. Valle, V. McKusick, J. Stanbury, J. Wyngaarten, D. Frederickson, J. Goldstein, and M. Brown. 1990. *Introduction to Human Biochemical and Molecular Genetics.* McGraw-Hill: New York (179 pp).

Bernstam, Victor A. 1992. *Handbook of Gene Level Diagnostics.* CRC Press: Boca Raton, FL (695 pp).

Gelehrter, T. D. and F. S. Collins. 1990. *Principles of Medical Genetics.* Williams and Wilkins: Baltimore, MD (324 pp).

McKusick, V. 1992. *Medelian Inheritance in Man.* Tenth edition. Johns Hopkins University Press: Baltimore, MD (2 volumes).

Read, A. 1989. *Medical Genetics: An Illustrated Outline.* J. B. Lippincott: Philadelphia, PA (152 pp).

Books That Discuss the Human Genome Project

Kevles, D. and L. Hood, Editors. 1992. *The Code of Codes.* Harvard University Press: Cambridge, MA (384 pp).

Lee, T. F. 1991. *The Human Genome Project.* Plenum Press: New York (332 pp).

Wills, C. 1991. *Exons, Introns, and Talking Genes.* Basic Books: New York (368 pp).

Additional Reading for Individual Chapters

Chapter One

Andrews, L. B., J. G. Fullarton, N. A. Holtzman, and A. G. Motulsky, eds. 1994. *Assessing Genetic Risks: Implications for Health and Social Policy.* National Academy Press, Washington, D.C. (338 pp).

Halley, D., J. Bijman, H. deJonge, M. Sinaasapple, H. Neijens, and M. Niermeijer. 1990. The cystic fibrosis defect approached from different angles—New perspectives on the gene, the chloride channel, diagnosis, and therapy. *Eur. J. Pediatrics* 149: 670–677.

McPherson, M. and R. Dormer. 1991. Molecular and cellular biology of cystic fibrosis. *Molec. Aspects Med.* 12: 1–81.

O'Loughlin, E. 1990. Cystic fibrosis: An inborn error of cellular electolyte transport? *J. Paediatr. Child Health* 26: 120–131.

Welsh, M. J. and A. E. Smith. 1993. Molecular Mechanisms of CFTR chloride channel dysfunction in cystic fibrosis. *Cell* 73: 1251–1254.

Chapter Two

Cairns, J., G. Stent, and J. Watson. 1966. *Phage and the Origins of Molecular Biology.* Cold Spring Harbor Laboratory of Quantitative Biology: Cold Spring Harbor, NY (340 pp).

Drlica, K. 1992. *Understanding DNA and Gene Cloning: A Guide for the Curious.* John Wiley & Sons, Inc.: New York (240 pp).

Fischer, E. P. and C. Lipsom. 1988. *Thinking About Science.* W. W. Norton and Co.: New York (334 pp).

Schroedinger, E. 1944. *What Is Life?* Cambridge University Press: Cambridge, England (99 pp).

Watson, J. D. 1980. *The Double Helix: A Personal Account of the Discovery of the Structure of DNA,* edited by G. Stent. W.W. Norton: New York (298 pp).

Chapter Three

Crick, F. 1988. *What Mad Pursuit.* Basic Books: New York (182 pp).

Drlica, K. 1992. *Understanding DNA and Gene Cloning: A Guide for the Curious.* John Wiley & Sons, Inc., New York (240 pp).

DOUBLE-EDGED SWORD

Kornberg, A. 1989. *For the Love of Enzymes*. Harvard University Press: Cambridge, MA (336 pp).

Stent, G. 1969. *The Coming of the Golden Age*. The Natural History Press: Garden City, NY (146 pp).

Chapter Four

Bishop, J. and M. Waldholz. 1990. *Genome*. Simon and Schuster: New York (352 pp).

Goodfellow, P. 1993. Planting alfalfa and cloning the Huntington's disease gene. *Cell* 72: 817–818.

Gusella, J., N. Wexler, P. Conneally, S. Naylor, M. Anderson, R. Tanzi, P. Watkins, K. Ottina, M. Wallace, A. Sakaguchi, A. Young, I. Shoulson, E. Bonilla, and J. Martin. 1983. A polymorphic DNA marker genetically linked to Huntington's disease. *Nature* 306: 234–238.

Hayden, M. 1981. *Huntington's Chorea*. Springer-Verlag: Berlin (192 pp).

The Huntington's Disease Collaborative Research Group. 1993. A novel gene containing a trinucleotide repeat that is expanded and unstable on Huntington's disease chromosomes. *Cell* 72: 971–983.

Chapter Five

Anderson, W. F. 1992. Human Gene Therapy. *Science* 256: 808–813.

Gordon, J. W. 1990. Micromanipulation of embryos and germ cells: an approach to gene therapy. *Am. J. Med. Genetics* 35: 206–214.

Grossman, M., S. Raper, K. Kozarsky, E. Stein, J. Engelhardt, D. Muller, P. Lupien, and J. Wilson. 1994. Successful *ex vivo* gene therapy directed to liver in a patient with familial hypercholesterolemia. *Nature Genetics* 6: 335–341.

Kohn, D., W. F. Anderson, and R. M. Blaese. 1989. Gene therapy for genetic diseases. *Cancer Investigation* 7: 179–192.

McLachlin, J., K. Cornetta, M. Eglitis, and W. F. Anderson. 1990. Retroviral-mediated gene transfer. *Progress in Nucleic Acids Research* 38: 91–135.

Verma, I. 1990. Gene Therapy. *Scientific American*, November: 68–84.

Additional Reading

Chapter Six

1991. Unreported findings shed new light on HIV dental case. *AIDS Alert* 6: 121–144.

Drlica, K. 1992. *Understanding DNA and Gene Cloning*. John Wiley & Sons: New York (240 pp).

Hazeltine, W. and F. Wong-Staal. 1988. The Molecular Biology of the AIDS Virus, *Scientific American* October: 52–62.

Chapter Seven

Chesney, P. J. 1989. Clinical aspects and spectrum of illness of toxic shock syndrome: Overview. *Reviews of Infectious Diseases* 11: S1-S7.

Kreiswirth, B., G. Kravitz, P. Schlievert, and R. Novick. 1986. Nosocomial transmission of a strain of Staphylococcus aureus causing toxic shock syndrome. *Annals of Internal Medicine* 105: 704–707.

Palca, J. 1992. The case of the Florida dentist. *Science* 255: 392–395.

Chapter Eight

State of Texas v. David Hicks (87th Judicial District Court, Freestone County, TX. Trial record, 1989.

Billings, P. 1992. *DNA on Trial*. Cold Spring Harbor Press: Cold Spring Harbor, NY (154 pp).

Chakraborty, R. and K. Kidd. 1991. The utility of DNA typing in forensic work. *Science* 254: 1735–1739.

Lewontin, R. and D. Hartl. 1991. Population genetics in forensic DNA typing. *Science* 254: 1745–1750.

Neufeld, P. and N. Colman. 1990. When science takes the witness stand. *Scientific American* 262 (May): 46–53.

Roberts, L. 1991. Fight erupts over DNA fingerprinting. *Science* 254: 1721–1723.

Chapter Nine

Frasier, S. D. 1983. Human pituitary growth hormone (hGH) therapy in growth hormone deficiency. *Endocrine Reviews* 4: 155–170.

Goeddel, D., H. Heyneker, T. Hozumi, R. Arentzen, K. Itakura, D. Yansura, M. Ross, G. Miozzari, R. Crea, and P. Seeburg. 1979.

Direct expression in *Escherichia coli* of a DNA sequence coding for human growth hormone. *Nature* 281: 544–548.

Chapter Ten

Bishop, J. and M. Waldholz. 1990. *Genome*. Simon and Schuster: New York (352 pp).

Gwyne, J. 1989. High-density Lipoprotein Cholesterol Levels as a Marker of Reverse Cholesterol Transport. *American Journal of Cardiology* 64: 10G–17G.

Johnson, C. and K. Edleman. 1992. Malignant hyperthermia: A review. *J. Perinatology* 12: 61–71.

McCarthy, T., Healy, J., Lehane, M., and J. Heffron. 1990. Recent developments in the molecular genetics of malignant hyperthermia: implications for future diagnosis at the DNA level. *Acta Anaesthesiol. Belg.* 41: 107–112.

Reichl, D. and N. Miller. 1989. Pathophysiology of Reverse Cholesterol Transport. *Arteriosclerosis* 9: 785–797.

Wingerson, L. 1990. *Mapping Our Genes*. Dutton: New York (338 pp).

Chapter Eleven

Potential use of live viral and bacterial vectors for vaccines. WHO Meeting, 1990. Geneva, 19–22 June, 1989. *Vaccine* 8: 425–437.

Andrews, L. B., J. G. Fullarton, N. A. Holtzman, and A. G. Motulsky, eds. 1994. *Assessing Genetic Risks: Implications for Health and Social Policy*. National Academy Press, Washington, D.C. (338 pp).

Duster, T. 1990. *Backdoor to Eugenics*. Routledge: New York (193 pp).

Primakoff, P., W. Lathrop, L. Woolman, A. Cowan, and D. Myles. 1988. Fully effective contraception in male and female guinea pigs immunized with the sperm protein PH-20. *Nature* 335: 543–546.

———————————— † ————————————

More About DNA and RNA

Most of this book has focused on how genetic information, in terms of nucleotide sequence, can influence our lives. In general, using this information requires the assistance of experts. There are, however, aspects of DNA and RNA biology that allow one to make common-sense judgments. For example, it has been known for many years that ultraviolet light can cause mutations in DNA. Since ultraviolet light is a component of sunlight, it was obvious that sunlight could cause serious skin problems long before clinical studies related skin cancer to sun exposure. This appendix introduces additional information about nucleic acids to enhance your ability to evaluate and act on future developments in molecular genetics.

DNA Structure

As pointed out in Chapter 2, DNA is the molecule that stores hereditary information. To do this, DNA must be capable of replication (duplication) so every cell gets a copy of the information, it must be chemically stable so errors are not easily introduced, and its genes must be recognized by other molecules so specific bits of stored information are available to particular cell types. How these requirements are met has been revealed in large measure by the study of DNA structure. Knowledge of DNA structure also allows us to guess about how agents in the environment can interfere with the activities of DNA.

157

DOUBLE-EDGED SWORD

Like other large biological molecules, DNA is composed of many smaller molecules (subunits) linked together. In DNA, the subunits are called nucleotides, and they are connected like beads in a necklace. Under most conditions, DNA is composed of two strands (some viruses have only one strand). Double-strandedness makes it difficult for an error to persist, since the information (nucleotide sequence) in one strand can be used to correct an error in the other. Nevertheless, our environment contains very potent agents (mutagens) that can change the information in DNA. Some of these agents, such as X-rays, can even break DNA. Breaks that occur in only one strand are readily repaired, since the ends are held close together by the interaction of the broken strand with the intact one. Double-strand breaks, however, allow the DNA ends to move apart, and repair is difficult. Double-strand breaks can lead to chromosomal disorders (see Appendix III).

A small region of a DNA molecule is depicted in Figure AII.1. Each nucleotide is itself composed of three types of chemical unit: a nucleic base, a sugar, and a phosphate. The bases are flat, ringlike structures formed by the joining of nitrogen and carbon atoms. Two of the bases are single rings, and two are double rings. Oxygen, hydrogen, nitrogen, and carbon atoms decorate the rings and make the bases chemically distinct. In DNA, each base is attached to a five-carbon sugar called deoxyribose. Deoxyribose units form part of the backbone of DNA, alternating with phosphates (a phosphate is one phosphorus atom attached to four oxygens). The sugar-phosphate backbones of the two DNA strands follow helical paths at the outer edge of an imaginary cylinder. The bases, each attached to a sugar, point toward the center of the cylinder with their flat sides perpendicular to the long axis of the helix, like steps in a spiral staircase.

The two strands of DNA are held together by weak attractive forces. One type arises from hydrogen bonds (represented by dotted lines in Figure AII.1) that form between bases in opposite strands. Another type comes from the tendency of the bases to stick together, much like oil droplets coalesce in water (one can imagine that water molecules surrounding DNA force the bases, and thus the strands, together). The flattened bases also stack on one another, minimizing their contact with water.

More About DNA and RNA

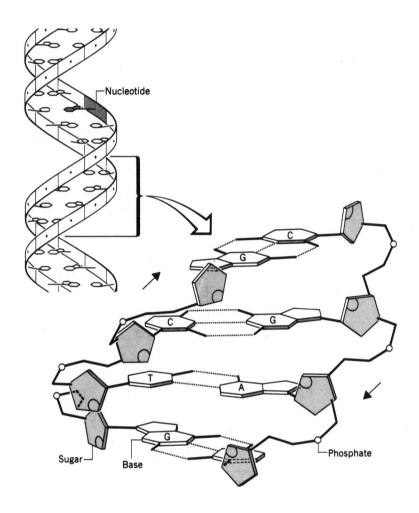

Figure AII.1. DNA structure.

A region of a DNA molecule is enlarged to show the directionality of the backbone and the hydrogen bonds that contribute to base-pairing specificity. Figure from *Understanding DNA and Gene Cloning* by Karl Drlica, John Wiley & Sons, New York.

The attractive forces are strongest when the bases in one strand align perfectly with those in the other. This is the molecular foundation for all of the DNA-based genetic tests—experimental conditions can be set up in which only perfectly matched sequences will pair and form double-stranded molecules. This allows us to ask whether a particular DNA strand has the same genetic information as another DNA strand.

DNA is not a smooth cylinder: it contains two grooves. The larger groove, called the major groove, is deep and wide while the smaller, minor groove is narrow and shallow. Many proteins that regulate the activities of DNA are currently thought to bind to one face of the DNA and form contacts with parts of DNA within the major groove. In a sense, these proteins dock on one side of the DNA. Many of these proteins recognize and bind to specific nucleotide sequences in DNA.

Careful inspection of the structure of nucleotides (base plus sugar plus phosphate) reveals that they are asymmetric: one end is different from the other. When nucleotides are joined to form DNA, the resulting DNA strand is like a chain of elephants connected trunk to tail. Like a chain of elephants, the DNA strands have distinct left and right ends (these ends are called 5′ and 3′, numbers that refer to specific carbon atoms in the sugar (deoxyribose)). An important feature of DNA is that the two strands of a double helix run in opposite directions (one chain of elephants is headed right and the other left). Many proteins that interact with DNA recognize the directionality. Those that must move along the DNA then "know" which way to move.

Complementary base-pairing

One of the most important concepts to emerge from understanding the structure of DNA is complementary base-pairing. In double-stranded DNA, an A in one strand always pairs with a T in the other strand and a G always pairs with a C. The principle is illustrated in Figure AII.2. This rule allows genetic information to be copied accurately during DNA replication, since each new strand must be complementary to each old one. The same rule applies to DNA•RNA hybrids (one strand is DNA and the other is RNA). RNA is similar to DNA except that U in RNA replaces T in DNA. Thus information in DNA can be converted to an RNA form by making a new strand of RNA (messenger

(a) **Structural formulas**

(b) **Prongs and sockets**

Figure AII.2. Complementary base-pairing.

(a) Structural formulas for two base pairs, the thymine:adenine (TA) and cytosine:guanine (CG) base pairs. The bases are flat structures composed of hydrogen, carbon, nitrogen, and oxygen atoms. The solid lines represent covalent bonds between these atoms. Arrows indicate where the bases attach to sugars in DNA. The dotted lines are hydrogen bonds, weak attractive forces between hydrogen and either nitrogen or oxygen. Notice that there are two hydrogen bonds between adenine and thymine, and three between guanine and cystosine. The difference in hydrogen bonding is part of the structural explanation for complementary base-pairing.

(b) A prongs-and-sockets analogy for base pairing. The hydrogen atoms in each hydrogen bond are represented as prongs, and the oxygen or nitrogen atoms are depicted as sockets. The attractive forces are weak; consequently, perfect fits are required for base-pairing to occur. Figure from *Understanding DNA and Gene Cloning* by Karl Drlica, John Wiley & Sons, New York.

DOUBLE-EDGED SWORD

RNA) from information in one strand of DNA. This process is called transcription.

The complementary base-pairing rule also makes it possible for one nucleic acid strand (DNA or RNA) to recognize another and specifically bind to it, since complementary base pairs make the best fit between two strands. This is important for the proper transfer RNA molecules to recognize the appropriate codon in messenger RNA during protein synthesis. Complementary base pairing is also the key to using DNA probes to identify DNA fragments for DNA fingerprinting analyses and virtually all DNA-based genetic analyses.

Several factors are involved in complementary base-pairing. One is that a single-ring base (C, T, U) always pairs with a double-ring base (G, A). This keeps the diameter of the double helix roughly constant. The hydrogen bonds between the bases (see Figure AII.2) are also of major importance. A•T(U) and G•C base pairs generate the strongest hydrogen binding patterns.

Plasmids

Some small, circular DNA molecules, called plasmids, can live as parasites inside bacterial cells. Plasmids contain signals in their DNA that cause the machinery of the host to replicate the plasmid. Since plasmids are small and are maintained at high copy numbers, they are good vehicles for cloning genes. Plasmids also encode a few genes of their own, with at least one responsible for controlling the number of plasmid copies present. Other plasmid genes often confer antibiotic resistance. These genes allow the host bacterium to grow in the presence of particular antibiotics. Plasmids are one of the reasons why virtually every bacterial pathogen of human beings has forms resistant to antibiotics.

The medical importance of plasmids has recently been reemphasized by a study relating antibiotic resistance to dental fillings. It had long been known that some plasmids encode genes that detoxify mercury. These plasmids allow their host bacteria to grow in the presence of mercury, and thus those bacteria tend to dominate the population when mercury is around. Those plasmids also carry genes for antibiotic resistance; consequently, the presence of mercury leads to antibiotic resistance as well. One source of mercury in our bodies may be amalgam tooth fillings from which mercury leaches. Experiments with animals showed that amalgam fillings correlated with increased antibiotic resistance in intestinal bacteria. Dentists tell us that amalgam is by far the best choice for filling dental cavities; thus we may be faced with a new dilemma.

Since some plasmids can move from one bacterium to another, antibiotic resistance can spread from benign, or even helpful, bacteria to pathogens. Some plasmids can even integrate (insert) into chromosomal DNA, and when they excise, they sometimes carry bacterial genes. As they move from bacterium to bacterium, they can transfer genes.

Transposons

Plasmids are but one way in which short pieces of DNA move around. Another is by a process called transposition. The regions of DNA that move are called transposons (an example is shown in Figure AII.3). In bacteria, transposons are usually detected by the presence of genes that confer antibiotic resistance. Transposons also contain genes for proteins required for their movement to new locations. Some transposons leave the original location when they move, while others make a copy that inserts at the new location. Transposons are important because they

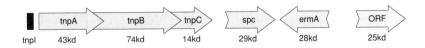

Figure AII.3.   A transposon found in staphylococcus DNA.

A transposon is a mobile element found inserted in DNA. It usually contains at least one gene whose product is required for the movement of the transposon (for the transposon shown, Tn554, there are three genes involved in transposition called *tnpA, tnpB,* and *tnpC*) and a gene that confers antibiotic resistance (in this case, there are two genes: *spc,* which gives resistance to spectinomycin, and *ermA,* which encodes erythromycin resistance). The arrows indicate the direction of transcription for each gene. The function of ORF (open reading frame) is unknown. The diagram was provided by Ellen Murphy, Public Health Research Institute.

provide sequence variability in DNA as they move around in it. They may also be important in moving drug resistance from one bacterium to another, since they can hop between chromosomes and plasmids.

RNA

RNA is similar to DNA, but it is much shorter and generally has only one strand. The sugar of RNA (ribose) has one more oxygen atom than the sugar of DNA, and that makes it more sensitive than DNA to breakage. The single-stranded nature of RNA allows it to fold upon itself more easily than DNA, and that enables RNA to adopt many different structures and to perform a variety of functions in cells. Among the known roles for RNA are storage of genetic information, movement of genetic information from DNA to cellular locations where the information is translated into protein, structure for cellular components such as ribosomes (workbenches where proteins are made), serving as adapters for aligning amino acids in the correct order when proteins are made, and acceleration of certain chemical reactions (some RNA molecules behave like enzymatic catalysts).

Storage of information as RNA is common among viruses. In some, the genome is single stranded, while in others it is double stranded. To replicate genomic RNA, single-stranded RNA viruses make a complementary copy of their RNA, and that copy is then used as a template for synthesizing the genomic RNA. In a few cases, RNA is converted to DNA before replication. This process, called reverse transcription, is the hallmark of retroviruses such as the one that causes AIDS. The DNA copy inserts into a host chromosome, and from there it encodes new viral RNA.

To access information in DNA, an enzyme called RNA polymerase attaches to DNA next to a gene and forces the two DNA strands apart. It then passes along the gene as it makes messenger RNA, using one of the DNA strands as a template. There are specific nucleotide sequences in DNA that indicate where synthesis of a particular messenger RNA is to start and other sequences that indicate where synthesis is to stop.

After messenger RNA is made, it attaches to ribosomes, which are themselves combinations of RNA molecules and specific protein molecules. There the information in messenger RNA is translated into protein. The translation process involves alignment of specific amino acids in the order specified by nucleotides in messenger RNA. The process aligns, without error, twenty different amino acids in chains that are sometimes one thousand amino acids long. Alignment is facilitated by short RNA molecules called transfer RNAs. Cells contain one or more specific transfer RNA types for each amino acid type. One end of the transfer RNA attaches to the amino acid, and the other end recognizes a specific three-nucleotide codon in messenger RNA. This places the amino acids in the correct order, and they are joined to form the new protein while still on the ribosome. The messenger RNA contains a specific codon that specifies where to start the protein chain as well as a stop codon to indicate the end of the protein.

Control of Gene Expression

At the 5′, upstream end of genes are nucleotide sequences that act as control regions determining when messenger RNA is to be made from the gene. As pointed out above, the enzyme that makes RNA from DNA is called RNA polymerase, and it recognizes and binds to a region of DNA called a promoter. Sometimes this region also contains a site for attachment of another protein that prevents RNA polymerase from binding. Such a protein is called a repressor, and it acts as a negative regulator. Still other proteins enhance the binding of RNA polymerase and thus the formation of messenger RNA. Regions where these proteins bind are often called enhancers. Elaborate networks of specific binding sites and proteins (transcription factors) exist to make RNA from genes in the correct amounts at the correct times in the correct cell types.

Regulation can also occur after messenger RNA has been made. One example occurs in egg cells, which are often full of untranslated messenger RNA. As soon as fertilization occurs, these messengers are quickly translated into protein so the developing embryo can grow quickly. Another example occurs with proteins that have a tail that keeps them inactive. When this tail is clipped off, the protein becomes active.

Common types of mutation

Occasionally errors are made during DNA replication, and if not corrected, the errors may be passed on to the next generation of cells. Changes in genetic information can have serious consequences; several examples are illustrated in Figure AII.4. In an offspring, a single nucleo-

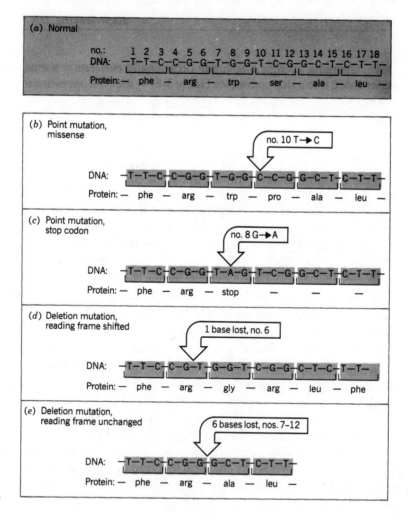

tide pair might be changed in the gene; for example, an AT pair might be converted to a GC, CG, or TA pair. Consequently, an incorrect amino acid may be inserted into the protein. Changing a single nucleotide may also convert a normal triplet codon into a stop codon; then the protein synthesis machinery prematurely stops making the protein. If such a mutation occurred near the beginning of the gene, only a small fragment of the protein could be made, with obviously serious effects. In another type of mutation a nucleotide is lost (or added) during replication; that is, the replication machinery skips (or adds) a "letter." In this case, the information is thrown out of the correct reading frame, and many incorrect amino acids are placed in the protein. This is called a frameshift mutation. In still other cases, a large stretch of DNA is lost. This is called a deletion mutation.

Within this framework, any chemical that converts one nucleotide "letter" to another in the old DNA strand is capable of "tricking" the replication machinery into inserting the wrong nucleotide into the new

Figure AII.4.   Common types of mutation.

(a) A normal nucleotide sequence for one strand of DNA codes for a protein having the six amino acids listed (phe, arg, trp, etc.). The codons for the amino acids are indicated by brackets in the DNA above the respective amino acids.

(b) If a T is changed to a C (arrow), the resulting mutant protein has a proline where serine is normally located. This type of change is called a missense mutation.

(c) If a G is changed to an A as in the example shown (arrow), a stop (nonsense) codon is created and protein synthesis halts. This change is called a nonsense mutation, and the mutant protein is shorter than the normal protein.

(d) Deletion of one base throws the reading frame out of register (frameshift mutation), and incorrect amino acids (gly, arg, leu) occur in the mutant protein.

(e) Removal of six bases produces a deletion mutation and a protein missing two internal amino acids (trp, ser). Figure from *Understanding DNA and Gene Cloning* by Karl Drlica, John Wiley & Sons, New York.

DOUBLE-EDGED SWORD

DNA strand as it is made. Chemicals that alter the information in DNA are called mutagens. Chemical mutagens, which are everywhere in our environment, and physical factors such as ultraviolet light, have become a major threat to human health. There is little doubt that both can lead to certain types of cancer. Household items, food additives, and pesticides are now being tested for their ability to cause mutations using bacteria. The test is based on our ability to easily detect the creation of bacterial mutants by whether they grow on agar plates containing particular nutrients.

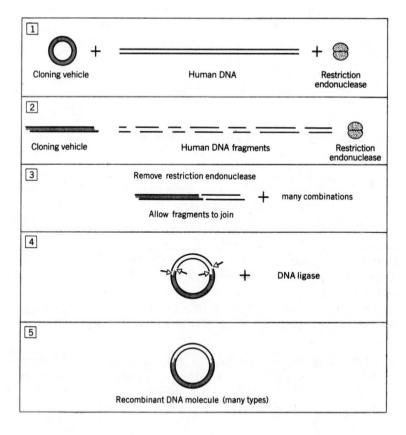

More About DNA and RNA

Gene cloning

Gene cloning is a way to obtain many copies of a specific region of DNA in a pure form. Developed in the mid-1970s, it has been the major technique driving the genetic revolution. The process involves a number of steps: (1) many identical DNA molecules are cut into many discrete pieces with a restriction endonuclease; (2) the resulting fragments are attached individually to infectious DNA molecules (plasmids or viral DNA), creating recombinant DNA molecules (Figure AII.5); (3) the recombinant DNA molecules are placed inside single-celled microorganisms such as bacteria or yeast so each cell receives only one recombinant DNA molecule; (4) the infected microorganisms are spread on a solid surface so individual cells separate from each other; (5) the separated cells each grow into a clone or colony of identical cells; (6) the colony containing the DNA fragment of interest is identified from among millions that lack that DNA piece; and (7) that colony is cultured to generate many billions of identical microorganisms, each containing

Figure AII.5.   General scheme for forming recombinant DNA molecules.

(1) Circular plasmid DNA (cloning vehicle), human DNA, and a restriction endonuclease are mixed.

(2) Both DNAs are cut, producing a linear plasmid and many human DNA fragments. In this example, all DNAs in the mixture have complementary, sticky ends.

(3) Occasionally a human DNA fragment will attach to one end of the plasmid. Many combinations form because many different types of human DNA fragments can join to the plasmid.

(4) Eventually both ends of the human DNA fragment will have attached to the respective, corresponding ends of the plasmid. When DNA ligase is added, the discontinuities in the DNA strands (arrows) will be sealed, producing a circular recombinant DNA molecule with no breaks in the DNA strands.

(5) The ligation of different human DNA fragments to plasmids causes the formation of many types of recombinant DNA molecule. Figure from *Understanding DNA and Gene Cloning* by Karl Drlica, John Wiley & Sons, New York.

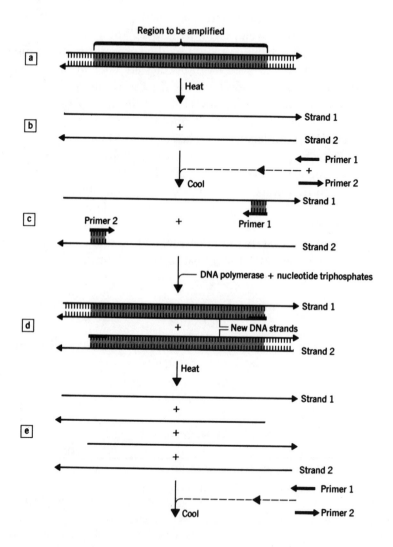

Fig. AII.6 continued on p. 172

More About DNA and RNA

the DNA fragment of interest. From this culture, large quantities of the recombinant DNA molecule can be isolated. The recombinant DNA can then be cut with a restriction endonuclease, and the fragment of interest can be separated from the cloning vehicle by gel electrophoresis for further study.

Polymerase chain reaction (PCR)

In the process called the polymerase chain reaction (Figure AII.6), a specific segment of DNA can be enriched by more than one hundred

Figure AII.6. Amplification of DNA by the polymerase chain reaction.

(a) A double-stranded DNA molecule is shown which contains a specific region of interest (double-strandedness is indicated by shaded areas, and the arrowhead on each strand represents the 3′ end). (b) Heating the DNA causes the two strands to separate. (c) When short primers complementary to regions on each of the two strands shown in (a) are added to the mixture and it is cooled, the primers hybridize to the two strands labeled 1 and 2. (d) The primers are extended by the addition of DNA polymerase and nucleoside triphosphates. At this stage the new DNA molecules have different lengths because of the asynchrony of the polymerization reaction. (e) When the mixture is heated, all of the DNA again will become single stranded. (f) Upon cooling, primers will hybridize to both old and new strands in the mixture. (g) DNA polymerase again extends the primers. In the cases where the templates are the strands made in step d, DNA synthesis stops when the polymerase reaches the end of the region to be amplified, for that is the end of the template. This produces a discrete DNA fragment (★) containing the nucleotide sequences of both primers and the DNA in between. (h) Subsequent heating, cooling, and polymerization increases the relative abundance of the discrete fragment (★). Heat-resistant DNA polymerase is used, so if enough polymerase, nucleoside triphosphates, and primers are added at the beginning of the reaction, the process consists only of heating and cooling steps. Figure from *Understanding DNA and Gene Cloning* by Karl Drlica, John Wiley & Sons, New York.

DOUBLE-EDGED SWORD

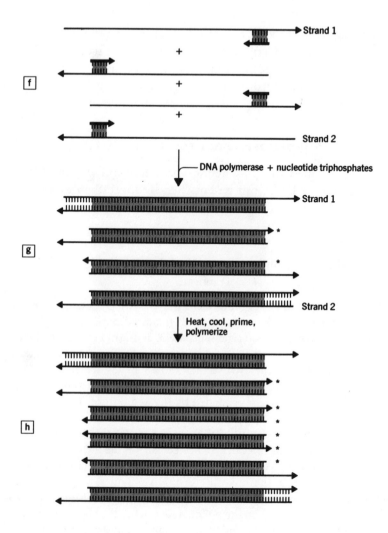

thousand times relative to nearby nucleotide sequences. The method is based on the observation that DNA polymerase can replicate a single-stranded DNA template, but only if a complementary fragment of DNA, a primer, is present to provide an end to which the polymerase can add nucleotides. As a result, DNA polymerase always begins replicating a

DNA template at a position defined by a primer forming base pairs with the template strand. When two strands of a long, double-stranded DNA molecule are separated and a short primer is hybridized to one strand at a specific spot (Figure AII.6c), DNA synthesis will always begin at the spot. Synthesis will proceed in only one direction from the primer because of the directionality of polymerase movement along DNA. A different primer can be placed on the second strand so that the region between the primers, which are at high concentration in the mixture, will be synthesized (Figure AII.6d). After DNA synthesis occurs, the mixture is heated to produce single-stranded molecules from the double-stranded ones. Upon cooling, the primers hybridize to the new DNA as well as to the original strands. Another round of DNA synthesis generates discrete DNA fragments that include the sequences of both primers and the DNA in between (Figure AII.6g). The cycle of heating, cooling, and polymerization is repeated on the order of thirty times, and with each cycle the discrete fragment increases in abundance. Because a DNA polymerase is used which can withstand heating, no reagents need be added between cycles. That makes the process easy to automate. Within three hours it is possible to obtain a specific fragment of DNA that can be easily purified. The limiting factor is knowing enough about the nucleotide sequence to generate the primers.

One major application of gene amplification is genetic testing. The amplification process generates a specific DNA fragment whose nucleotide sequence can be easily determined, so the presence of mutations associated with genetic diseases can be detected. The method can also be used to detect viral diseases such as AIDS. The nucleotide sequence of the viral genome is known, and if it is present in even 1 in 1,000 human cells, it can be detected by amplification. A third application concerns the identification of people. Amplification and nucleotide sequence determination of particularly variable genes is a variation on the hybridization method described in Chapter 8. The method is so sensitive that a single human hair cell may provide enough material for analysis. Amplification will become increasingly important in criminal cases, for it will enable law enforcement agencies to clearly identify people from hair or skin fragments found at the scene of a crime.

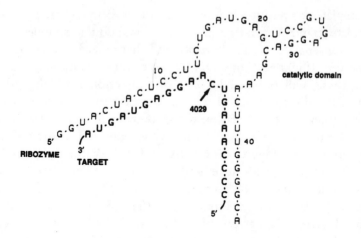

Figure AII.7. Binding of a hammerhead ribozyme to its target.

A ribozyme is shown aligned with its target, a portion of an HIV-1 messenger RNA. The arrow indicates the C of the GUC target. The ribozyme is composed of a catalytic domain that does the cutting and two flanking regions that recognize and attach to the target according to the complementary base-pairing rules (see Figure AII.2). Figure adapted from M. Sioud and K. Drlica, Proc. Natl. Acad. Sci. U.S.A. 88: 7303–7307 (1991).

Therapeutic Nucleic Acids

One type of small nucleic acid being studied for its therapeutic potential is called antisense RNA. Antisense RNAs are designed to form complementary base pairs with specific messenger RNA molecules that encode harmful proteins. When an antisense RNA is attached to a messenger RNA, that messenger is not translated effectively into protein. Even more potent is antisense DNA. When this single-stranded DNA binds to messenger RNA, it is digested by an enzyme that cleaves RNA only when base-paired to DNA.

A ribozyme is another type of small RNA that has therapeutic potential. These RNA molecules can specifically recognize other RNA molecules and then cut them. Ribozymes were originally found through their ability to cut themselves, and once the crucial nucleotide sequences were identified, ribozymes were designed to specifically cut other RNA molecules. For the type called a hammerhead, the target RNA need only contain one of several nucleotide triplets (such as GUC). The ribozyme is then made so it will form complementary base pairs with the region of the target RNA around the sensitive triplet. The ribozyme then recognizes its specific target by complementary base-pairing, binds to it, and cleaves it next to the triplet (Figure AII.7). Since many RNA molecules carry a suitable triplet that makes them susceptible to ribozyme attack, the range of potential targets of a hammerhead ribozyme is immense. By making the recognition sequences fairly long, there is little chance that the ribozyme will make a mistake and cut the wrong target. These RNA molecules are being tested for their antiviral activity and for use in gene therapy to block harmful genes.

To construct a ribozyme directed against any particular target, one must first synthesize a short DNA molecule to act as a template for making the desired ribozyme RNA molecule. This DNA molecule represents a ribozyme gene, which can then be placed inside living cells. There it specifies production of a ribozyme that will bind to a particular target RNA and cleave it.

Antisense RNA and ribozymes provide us with specific ways to inactivate genes in a controlled fashion and in principle make it unnecessary to replace harmful genes, a process for which there is no simple method. In addition, most infectious diseases caused by viruses should be susceptible to ribozyme therapy. In these cases the ribozyme would be designed to cut the messenger RNA from a viral gene essential for producing new viruses.

---- † ----

Family Analysis

Knowing as much as possible about your genes is one way to prepare for future developments in molecular genetics. Currently, that knowledge can best be obtained by collecting medical information about your family. Even if you don't expect to find one of the obvious disorders, your collection could be valuable for your children and grandchildren; they will have access to many more DNA tests than you have. This appendix provides pointers for constructing family trees and describes patterns of inheritance. It also includes a brief description of DNA extraction methods, since some families are already storing DNA samples for future analysis.

Two notes of caution should be considered before beginning family medical analysis. First, uncovering genetic information can be very unsettling. You may turn up a family disorder that has no cure, and some of your relatives might not want to know their future. Be careful about communicating your conclusions, especially before they are confirmed by a professional. Second, information you uncover can be potentially harmful to your family in terms of insurance, employment, and adoption (see Appendix IV). Keep the information confidential.

Constructing a Family Tree

Your first task is to identify hereditary problems and to define their patterns of inheritance. Then you'll be able to estimate the probability

Family Analysis

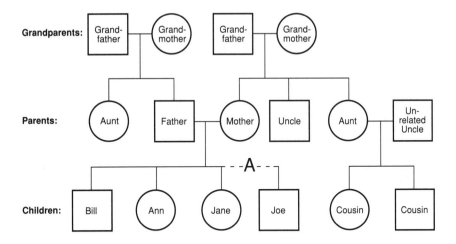

Figure AIII.1. Diagram of a family tree.

Males are represented by squares, and females by circles. Legal relatives that do not share the same genetic background should be noted. In the figure, the letter *A* denotes that Joe was adopted.

that a particular disorder will affect you or your children. One strategy for seeing relationships and tracking a disorder is to construct a diagram called a family tree. An example of a small portion is shown in Figure AIII.1. In this diagram, males are indicated by squares and females by circles (as discussed later, sex can be important in the pattern of inheritance). Children are usually listed from left to right in order of birth. Four children are named in the example. Joe was adopted, and this is noted in the tree. Special designations are important for such cases, which include adoption, artificial insemination, surrogate parentage, and extramarital parentage. In these instances, the children do not have the same genetic background as their siblings, and conclusions based on their features could be misleading.

As you collect medical information, add it to the tree. In the beginning, you will be looking for anything abnormal (for list, see Table AIII.1). At a minimum, you will want to record the age and cause of

Table AIII.1. Disorders That Are Often Inherited

autoimmune disease	immune deficiency
bacterial infections,	lactose digestion, deficient
chronic	life span reduced
blood coagulation	male infertility
impaired	mental retardation
bone fragility	miscarriages
cancer, predisposition to	muscle cramps, excessive
cataracts	muscle weakness in adults
cleft lip and palate	neurologic abnormalities
diabetes	retinal tumors
exercise intolerance	schizophrenia
gout	sexual development delayed
hearing loss	solar skin damage, susceptibility to
heart attacks	squamous cell carcinoma of skin,
heart disease, congenital	susceptibility to
hypertension	stature, very short
hypothyroidism, goiter	visual acuity reduced

death for as many family members as possible. Beyond that, it is probably best to assume that you do not know what to look for. Thus you will need to record everything. If you have access to considerable information, it may be best to write the information on cards keyed to persons on the tree (suitable genealogical computer programs are available).

Some of the details you collect could have obvious importance. For example, if your father died young from lung cancer, your concern

Family Analysis

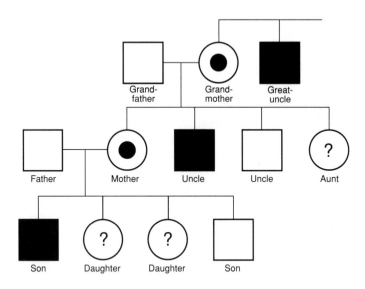

Figure AIII.2. Inheritance pattern of an X-linked recessive disorder.

Squares indicate males, circles females. Black symbols represent persons showing disease symptoms. Asymptomatic carriers are indicated by a dot in the circle; question marks refer to cases where carrier status is undetermined.

about a genetic predisposition to cancer would be very different if he were a cigarette-smoking coal miner than if he neither smoked nor lived where the air was polluted. Other factors will become significant only after further scientific research. For example, if your cousin had five miscarriages and if her husband had previously been sprayed with Agent Orange while a soldier in Vietnam, your interpretation would depend on whether Agent Orange is ever shown to cause chromosomal damage to developing sperm.

When you are ready to look for genetic patterns, you will probably need to simplify the information. One way is to prepare a separate copy of the tree for each malady. Then symbols, such as solid squares or circles, can be used to indicate persons affected by the disorder (see Figure AIII.2 for an example of a disorder affected by sex).

For some disorders, family tree analyses reveal very clear patterns of inheritance (discussed in the following sections). You may detect one of these patterns in your family. However, it is more likely that you will see something that seems to just "run in the family." Human beings are complex organisms, and few, if any, genes act on their own. Thus, for most disorders inheritance patterns are complicated. Even a dominant trait can be influenced by many factors in the genetic makeup of an affected individual. These ill-defined factors may cause the same genetic disorder to have different manifestations in different individuals, making it difficult to identify the disease in some generations. Moreover, families may differ with respect to the particular nucleotide alteration in the gene responsible for the disease, so the severity of the disease and the age of onset can also vary from family to family. Still another complication is that the same symptoms sometimes arise from different molecular defects. This can cause the inheritance patterns to differ among families. These types of complexities, plus possible effects from the environment, may make additional reading or professional advice important for interpreting family medical data.

Some of the data you collect may be inaccurate, due either to misdiagnosis or faulty memories of your sources. Often there will be little you can do to verify the accuracy of the information. However, you can record the source. At a later time you can decide whether to trust the data.

Single Gene Disorders: Autosomal Dominant

Autosomal refers to all chromosomes other than the sex chromosomes (X and Y), and dominant means that a trait is manifested even when only one of the two gene copies is abnormal. Consequently, autosomal dominant diseases occur equally in males and females, and both sexes pass the disease to their children. In the classic cases, every affected individual will have one parent who is also affected (for pattern of inheritance see Figure 4.1). Since these diseases are rare, an affected person usually has only one copy of the responsible gene in the abnormal state and usually mates with someone having two normal gene copies. Thus, the chance for a child to be afflicted is 50 percent, high enough odds for the disease to show up in every generation. However, serious dominant disorders, which tend to strike adults, can appear to skip a generation if an affected person dies of other causes before showing signs of the disorder. This feature is important for identifying unaffected branches in the tree (the disorder cannot be transmitted through a person who has normal genes, and such persons would define a normal branch). Huntington disease (Chapter 4) is a typical dominant disorder.

From a biochemical point of view, one can think of autosomal dominant disorders as arising from either insufficient activity of a normal gene product or creation of a harmful product. In the former situation, therapy may be possible by increasing the concentration of the product. Eliminating a harmful product may in the future be approached using antisense RNA or ribozyme therapies.

Single Gene Disorders: Autosomal Recessive

Autosomal recessive diseases are clinically apparent only when both copies of the responsible gene are mutant. By definition, these autosomal genes are not located on the sex chromosomes, so males and females are affected in equal proportions. Parents generally appear normal; only siblings are affected. Since the probability that a child will be affected is only 25 percent, multiple occurrences in a family may not be seen where small families are common. Consequently, occurrences of this type of disease will appear sporadic. Cystic fibrosis, PKU, and Tay-Sachs disease are examples of this type of disorder. Their inheritance pattern is shown in Figure 1.1.

The biochemical defects for many autosomal recessive diseases occur in enzymes. This may arise because enzymes often provide almost normal function when at only half the normal level, the level found in carriers of these diseases. Then the disease is seen only when both copies of the gene are knocked out. Recessive disorders are generally diagnosed in children, and there is hope that many can be cured by gene therapy that adds a normal copy of the gene to the affected person (see Chapter 5).

Single Gene Disorders: X-linked

X-linked disorders arise from genes located on the X chromosome. Since males have one X and one Y, while females have two X chromosomes, the clinical manifestations and inheritance patterns differ for males and females. An affected male can never pass the disorder to his sons because sons always receive the Y, not the X, from their fathers. But an affected father passes his only X chromosome to all of his daughters, so they will each receive the mutant gene. Three patterns of X-linked inheritance are discussed below.

Recessive X-linked disorders are expressed in males, since males have only one X chromosome and thus only one copy of that set of genes (see pattern in Figure AIII.2, which is characteristic for hemophilia). Mothers and sisters of affected boys will generally be carriers without symptoms, since they have two X chromosomes. When family trees are examined, the disorder will appear in maternal uncles of affected males and in male cousins descended from sisters of the mother. Sons of carrier mothers have a 50 percent chance of having the disorder. An affected female offspring would arise only when both of her X chromosomes carry the disorder. This could happen if an affected man marries a woman who is either affected or is a carrier. Color blindness is one of the X-linked recessive disorders common enough to show up in women.

Dominant X-linked disorders appear in females as well as in males. While an affected mother transmits the disorder to only half of her daughters or sons, an affected father transmits it to all of his daughters but to none of his sons. Consequently, females tend to display this type of disorder twice as often as males.

A third pattern arises when a dominant X-linked trait is lethal for all males, since there can be no carrier father to pass along a mutant X chromosome (all such males die before birth). Diagnostically, women carrying this type of disorder have a high rate of spontaneous abortion of male fetuses.

The analysis of X-linked disorders in females can be complicated by X-chromosome inactivation, a phenomenon in which one of the two X chromosomes is irreversibly inactivated in all somatic cells of female embryos. This means that genes on only one of the two X chromosomes are expressed. The complicating feature is that inactivation is random: in some cells the paternal X will be active, while in other cells the maternal X will be active. Once inactivation occurs, the same chromosome remains inactive in all cells arising from that embryonic cell. X-inactivation results in females that are genetic mosaics: their tissues contain clusters of cells (each cluster derived from a single embryonic cell) in which one or the other X chromosome is active. The random nature of X-inactivation means that different females will have different clusters of cells affected by the disorders. Consequently, females can vary considerably in the manifestation of X-linked disorders.

Multifactor Disorders

Many disorders "run in families" without a clearly predictable pattern. Among the more common diseases of this type are hypertension, gout, diabetes mellitus, and peptic ulcer disease. It is generally thought that these are multifactor disorders that arise when a particular combination of genes and environmental conditions is present. The idea has gained support from the estimate that almost 30 percent of normal genes in humans are polymorphic and normally carry slight variations. Thus, the level to which human beings are predisposed to a particular multifactor disorder will vary considerably among individuals, and the patterns of inheritance will be difficult to work out. Family histories will be especially important as our descendants struggle to understand and control multifactor genetic disease.

New Mutations

New mutations can occur in any generation and can be inherited from then on. It is estimated that about 1 in 100,000 individuals acquires a mutation in a given gene. However, many of these mutations have no harmful effect, so the frequency of clinically expressed, single-gene disorders from new mutations is much, much lower. Since new mutations causing recessive disorders are difficult to identify (carriers don't manifest the disease), the new disease may not appear for generations. But dominant disorders or X-linked ones do reveal new mutations often enough to be detected. In those cases, new mutations occur slightly more often with older fathers.

Chromosomal Disorders

Most of the time, chromosomes are diffuse, DNA-containing structures located in a large intracellular organelle called a nucleus. However, during a short period when cells divide, chromosomes condense enough to be easily distinguished by light microscopy. At that stage, a chromosome looks like a hot dog with a constriction. The location of this constriction is characteristic for each chromosome. Prior to cell division, the chromosomes duplicate, and for a while the two daughter chromosomes remain attached. This gives the chromosomes an X-like appearance when viewed with a microscope. Since each chromosome pair has a characteristic shape, the pairs can be easily identified (chromosomes also have characteristic staining patterns when treated with dyes). Examination of many, many cells has revealed that chromosomes are occasionally gained or lost, broken with loss of a piece, broken and rejoined to another chromosome, or deleted for one arm while duplicated for the other. Some of these events occur often enough to be correlated with particular diseases. For example, Down syndrome is associated with the presence of three rather than two copies of chromosome number 21.

Chromosomal disorders occur fairly often (as many as half of the recognized, first-trimester spontaneous abortions contain chromosomal aberrations). Fortunately, these disorders can be easily detected by examination of chromosomes in fetal cells; thus, parents can be advised to abort fetuses known to be affected.

A variety of factors influence the frequency and types of chromosomal disorders, most of which are undefined. It is clear, however, that these disorders increase with maternal age, making fetal chromosome examination highly advisable for older mothers. It is also likely that particular genes predispose some families to chromosomal disorders. Thus, in families where miscarriage early in pregnancy is common, even young pregnant women should have fetal cells examined.

DNA Isolation and Storage

Since genetic information is stored in DNA, some families with histories of complex genetic disease have opted to have DNA or cell samples prepared and stored. These families anticipate that future tests on their samples will benefit subsequent generations. In 1992 it was estimated that commercial biotechnology laboratories had about 10,000 samples stored, mostly under the guidance of geneticists. Before participating in this type of service, it is important to carefully examine issues such as who has access to your sample, how it will be used, and what will happen to it if the company disbands.

If you are not in a research study and you want to store your DNA, you may wish to consider keeping it yourself in a safe-deposit box. DNA is quite stable when dry. To provide a sense of how DNA is obtained, an isolation method is described below.

DNA is a long, stringy molecule that can be easily separated from most other cellular components. A common practice is to first treat a tissue sample with detergent to break open the cells. Detergent also strips proteins from the DNA. Sometimes salt is added to the sample to cause proteins to precipitate (clump). They can then be removed by centrifugation. Any remaining proteins can be broken down by treating the mixture with proteases (enzymes that specifically cut proteins). Sometimes the watery mixture is shaken with phenol, an oily compound that separates from water when allowed to stand. During shaking, many contaminants move into the phenol, and they can be removed as the phenol is separated from the DNA-containing solution. Addition of alcohol to a solution of DNA causes the DNA strands to precipitate. If the alcohol is gently layered on a concentrated DNA solution, the DNA precipitates at the boundary between the alcohol and the DNA solution. A glass rod can then be stuck into the mixture and rotated between the fingers. That winds precipitated DNA around the rod-like string. When the rod is lifted out of the solution, the DNA spooled onto it looks much like nasal mucus. The DNA can then be dried and stored. Precipitated DNA can also be driven into the bottom of a tube by centrifugal force. After removing the upper fluid, water can be added to the sample to dissolve the purified DNA. Although these steps may have to be repeated a few times, they are adequate for preparing DNA for analysis by electrophoresis.

By slightly modifying the methods sketched above, even amateurs can obtain DNA from tissues. The main problem is obtaining the human tissue or blood. In the future, DNA amplification methods such as PCR (Appendix II) will make it necessary to have only small samples of dried blood, perhaps stored on pieces of filter paper (already the military is storing dried blood samples this way for future identification purposes). Then it will be relatively easy for families to accumulate the raw materials needed for family analysis.

---------------------------------- † ----------------------------------

The Law and
Genetic Discrimination

Cases of genetic discrimination are being documented and reported in the scientific literature. In one study (P. Billings et al., *American Journal of Human Genetics* 50: 476–482, 1992), more than a thousand professionals working in human genetics were solicited for reports of cases of possible discrimination. About forty incidents emerged. Thirty-two involved insurance problems related to health, life, disability, mortgage, and auto coverage. These issues tended to arise when relocation or a change of employer required changes in existing policies. Seven cases of discrimination involved employment issues such as hiring, termination, promotion, and transfer. In two cases, couples were excluded from consideration by adoption agencies on the basis of a disease running in their families (the couples were healthy). At present, we don't know how frequently genetic discrimination occurs, since this study was not a survey.

Often discriminatory behavior is based on a misunderstanding of the disease. Some genetic diseases can be successfully treated, and carriers of a recessive disease are often perfectly normal. In some diseases, the symptoms are very mild. Consequently, blanket discrimination is medically unjustified. In cases such as Huntington disease, in which a person is simply at risk, discrimination is just a matter of playing the odds. These cases portend a serious problem for the application of genetics, since they could cause people to avoid treatment if that meant tainting their family medical record.

If you choose not to reveal genetic information about yourself, you can point out to employers that there are laws that protect against this type of discrimination. Existing legislation thought to be relevant to genetic discrimination is sketched below. For more details and references, readers are encouraged to see Natowicz et al., *American Journal of Human Genetics* 50: 465–475, 1992.

The Federal Rehabilitation Act of 1973 prohibits employment discrimination on the basis of handicap by employers that receive federal monies. The key aspect is that a handicapped person includes anyone who is *regarded* as having an impairment that substantially limits one or more major life activities. Myths and fears about disability and disease are as handicapping as the physical effects of actual impairment. Thus this act appears to apply to purely asymptomatic conditions. Under the act, employers are expected to provide accommodations for a handicap, which would include moving an employee who is particularly sensitive to particular chemicals due to a genetic condition. This law also prohibits preemployment questions about health or disability, and it allows an employer to require a medical examination only after a job offer has been made and only if all prospective employees must take the medical examination.

The Americans with Disabilities Act of 1990 is similar to the 1973 act, but it is not limited to employers receiving federal funds. This law allows the preemployment medical examination to include genetic tests, but the results can be used only to exclude an individual from employment if the genetic finding makes the person unfit for the job. However, an employer may provide discriminatory insurance coverage.

A few states have laws that specifically limit the use of genetic information by employers or insurers. For example, Maryland requires actuarial information for the use of genetic information by insurers, and several states prohibit discrimination against carriers of specific recessive diseases.

---------------------------------- † ----------------------------------

Where to Get Help

Genetic Services Providers

Medical geneticists and genetic counselors are involved in genetic education at various genetic centers across the country. Many centers have formal education programs, a few publish public education newsletters, and most can provide staff members as guest speakers. State governments have genetic services and education programs administered either by the state Maternal and Child Health (MCH) Program or the state Program for Children with Special Health Care Needs (CSHCN). Every state is involved in a regional genetics network. These networks improve communication and coordination of service delivery among states within each region, standardize laboratory procedures, provide information, and gather data. The networks publish newsletters that provide information about area workshops and conferences, including those that focus on human genetic education. Listed below are some of the providers (taken from *Human Genetics: A Resource Guide,* published by the National Center for Education in Maternal and Child Health).

Council of Regional Networks for Genetic Services
(CORN)
Arizona Department of Health Services
Division of Disease Prevention,
Office of Risk Assessment and Investigation
3008 North Third Street, Suite 101
Phoenix, AZ 85012
(602) 230-5868

DOUBLE-EDGED SWORD

Genetics Network of the Empire State
(GENES)
Laboratory of Human Genetics
Wadsworth Center for Laboratories and New York State Department
of Health
Empire State Plaza
P.O. Box 509
Albany, NY 12201-0509
(518) 474-7148

Great Lakes Regional Genetics Group
(GLaRGG)
(Indiana, Illinois, Michigan, Minnesota, Ohio, Wisconsin)
Genetic Diseases Section
Division of Maternal and Child Health
Indiana State Board of Health
1330 West Michigan Street
Indianapolis, IN 46206-1964
(317) 633-0644

Great Plains Genetics Service Network
(GPGSN)
(Arkansas, Iowa, Kansas, Missouri, Nebraska, Oklahoma, North
Dakota, South Dakota)
University of Iowa
Department of Pediatrics
Division of Medical Genetics
Iowa City, IA 52242
(319) 356-2674

Mid-Atlantic Regional Human Genetics Network
(MARHGN)
(Delaware, District of Columbia, Maryland, New Jersey,
Pennsylvania, Virginia, West Virginia)
University of Virginia
School of Medicine,
Division of Genetics
P.O. Box 386
Charlottesville, VA 22408
(804) 924-9477

Mountain States Regional Genetics Services Network
(MSRGSN)
(Arizona, Colorado, Montana, New Mexico, Utah, Wyoming)
Colorado Department of Health
Family Health Service Division
4210 East 11th Avenue
Denver, CO 80220
(303) 331-8376

New England Regional Genetics Group
(NERGG)
(Connecticut, Maine, Massachusetts, New Hampshire, Rhode Island,
Vermont)
36 Ledge Lane
P.O. Box 682
Gorham, ME 04038-0682
(207) 839-5324

Pacific Northwest Regional Genetics Group
(PacNoRGG)
(Alaska, Idaho, Oregon, Washington)
Oregon Health Sciences University
Child Development and Rehabilitation Center
P.O. Box 574
Portland, OR 97207
(503) 494-8342

Pacific Southwest Regional Genetics Network
(PSRGN)
(California, Hawaii, Nevada)
California Department of Health Services, Genetics Disease Branch
2151 Berkeley Way, Annex Four Berkeley, CA 94704
(415) 540-3288

Southeastern Regional Genetics Group
(SERGG)
(Alabama, Florida, Georgia, Kentuck, Louisiana, Mississippi, North
Carolina, South Carolina, Tennessee)
Emory University School of Medicine
2040 Ridgewood Drive
Atlanta, GA 30322
(404) 727-5840

Texas Genetics Network
(TEXGENE)
Texas Department of Health
1100 West 49th Street
Austin, TX 78756-3199
(512) 458-7700

Volunteer Organizations

Volunteer organizations can act as support groups for coping with the
medical and psychosocial impact of genetic conditions and birth defects.
Some of these organizations are listed below in groups according to
disease categories. The list has been taken from *A Guide to Selected
National Genetic Voluntary Organizations* (originally published by the
National Center for Education in Maternal and Child Health) and
updated according to the *Directory of National Genetic Voluntary Organizations* (published by the Alliance of Genetic Support Groups). A more
comprehensive list, as well as updated addresses and telephone numbers,
may be obtained from the Alliance of Genetic Support Groups (address
below).

General Groups

Alliance of Genetic Support Groups
35 Wisconsin Circle, Suite 440
Chevy Chase, MD 20815
(800) 336 GENE

March of Dimes Birth Defects Foundation
1275 Mamaroneck Ave.
White Plains, NY 10605
(914) 428-7100

National Association of Radiation Survivors
(NARS)
78 El Camino Road
Berkeley, CA 94705
(415) 658-6056, 652-4400, ext. 441

The National Organization for Rare Disorders
(NORD)
P.O. Box 8923
New Fairfield, CT 06812
(800) 999-NORD or (203) 746-6518

National Easter Seal Society
2023 West Ogden Ave.
Chicago, IL 60612
(312) 243-8400

National Foundation for Jewish Genetic Diseases, Inc.
250 Park Ave., Suite 1000
New York, NY 10177
(212) 371-1030

Sibling Information Network
University Affiliated Program on Developmental Disabilities
University of Connecticut
1776 Ellington Rd.
South Windsor, CT 06074
(203) 648-1205

TASH: The Association for Persons with Severe Handicaps
7010 Roosevelt Way, NE
Seattle, WA 98115
(206) 523-8446

Auditory Groups

Alexander Graham Bell Association for the Deaf
(AGBAD)
3417 Volta Place, NW
Washington, DC 20007
(202) 337-5220

American Society for Deaf Children
(ASDC)
814 Thayer Ave.
Silver Spring, MD 20910
(800) 942-2732

Cancer Groups

American Cancer Society, Inc.
1599 Clifton Rd., N.E.
Atlanta, GA 30329
(404) 329-7622

Candlelighters Childhood Cancer Foundation
7910 Woodmont Ave., Suite 460
Bethesda, MD
(301) 718-2686

Familial Polyposis Registry
Mt. Sinai Hospital
600 University Ave., Suite 1157
Toronto, Ontario M5G 1X5
CANADA
(416) 586-8334

GI Polyposis and Hereditary Colon Cancer Registry
The Moore Clinic
Johns Hopkins Hospital
600 North Wolfe St.
Baltimore, MD 21205
(301) 955-4040 or (301) 955-3875

Gilda Radner Familial Ovarian Cancer Registry
Elm and Carrolton St.
Buffalo, NY 14263
(800) 682-7426

Intestinal Multiple Polyposis and Colorectal Cancer
(IMPACC)
1008–101 Brinker Dr.
Hagerstown, MD 21740
(301) 791-7526

Leukemia Society of America, Inc.
733 Third Ave.
New York, NY 10017
(212) 573-8484

National Cancer Care Foundation (NCFF)
1180 Ave. of the Americas
New York, NY 10036
(212) 221-3300

Cardiovascular Group

Council on Cardiovascular Disease in the Young
American Heart Association National Center
7320 Greenville, Ave.
Dallas, TX 75231
(214) 373-6300

Chromosomal Groups

Association for Children with Down Syndrome, Inc.
(ACDS)
2616 Martin Ave.
Bellmore, Long Island, NY 11710
(516) 221-4700

5p-Society
11609 Oakmont
Overland Park, KS 66210
(913) 469-8900

Fragile X Support, Inc.
1380 Huntington Drive
Mundelein, IL 60060
(312) 680-3317

National Association for Down Syndrome
(NADS)
P.O. Box 4542
Oak Brook, IL 60521
(312) 325-9112

National Down Syndrome Congress
(NDSC)
1800 Dempster St.
Park Ridge, IL 60068-1146
(312) 823-7550

National Down Syndrome Society
(NDSS)
666 Broadway
New York, NY 10010
(800) 221-4602; (212) 460-9330

National Fragile X Foundation
1441 York St., Suite 215
Denver, CO 80206

Prader-Willi Syndrome Association
(PWSA)
6490 Excelsior Boulevard E-102
St. Louis Park, MN 55426
(612) 926-1947

Support Group for Monosomy 9P
43304 Kipton Nickel Plate Rd.
La Grange, OH 44050
(216) 775-4255

Turner's Syndrome Society of the U.S.
15500 Wayzata Blvd., #768-214
Wayzata, MN 55391
(612) 475-9944

Connective Tissue Groups

Ehlers-Danlos National Foundation
(EDNF)
PO Box 1212
Southgate, MI 48195
(313) 282-0180

National Marfan Foundation
(NMF)
382 Main St.
Port Washington, NY 11050
(516) 883-8712; (800) 862-7326

Craniofacial Groups

**American Cleft Palate Associaton (ACPA)/
The Cleft Palate Foundation**
1218 Grandview Ave.
Pittsburgh, PA 15211
(800) 24-CLEFT; (412) 481-1376

**FACES—The National Association for the Craniofacially
Handicapped**
P.O. Box 11082
Chattanooga, TN 37401
(615) 266-1632

National Foundation for Facial Reconstruction
317 East 34th Street
New York, NY 10016
(212) 263-6656

Prescription Parents, Inc.
P.O. Box 161
West Roxbury, MA 02132
(617) 527-0878

Developmental Disabilities Groups

Association for Children and Adults with Learning Disabilities, Inc.
(ACLD)
4156 Library Rd.
Pittsburgh, PA 15234
(412) 341-1515

Association for Retarded Citizens of the United States
(ARC)
500 East Border Street, #5-300
Arlington, TX 76010
(817) 261-6003

Autism Society of America
(ASA)
8601 Georgia Ave. #503
Silver Spring, MD 20910
(301) 565-0433

Center for Hyperactive Child Information, Inc.
(CHCI)
P.O. Box 66272
Washington, DC 20035-6272
(703) 920-7495

Cornelia de Lange Syndrome (CdLS) Foundation, Inc.
60 Dyer Avenue
Collinsville, CT 06022
(800) 223-8355; (203) 693-0159

**Laurence-Moon-Biedl Syndrome
(LMBS) Support Network**
76 Lincoln Ave.
Purchase, NY 10577
(914) 251-1163

Orton Dyslexia Society
(ODS)
Cluster Bldg., Suite 382
8600 La Salle Rd.
Baltimore, MD 21204
(410) 296-0232

Progeria International Registry
North Shore University Hospital
300 Community Dr.
Mahasset, NY 11030
(718) 494-5333

Rubenstein-Taybi Syndrome Parent Group
(RTS)
414 East Kansas
Smith Center, KS 66967
(913) 282-6237

Share and Care
1294 "S" Street
North Valley Stream, NY 11580
(516) 825-2284

United Cerebral Palsy Associations, Inc.
(UCPA)/UCP Research and Educational Foundation
1522 K St., NW, Suite 1112
Washington, DC 20005
(800) USA-5UCP

Gastrointestinal Groups

American Celiac Society
45 Gifford Ave.
Jersey City, NJ 07304
(201) 432-1207

American Liver Foundation
(ALF)
998 Pompton Ave.
Cedar Grove, NJ 07009
(201) 857-2626

Celiac Sprue Association/United States of America, Inc.
(CSA/USA)
P.O. Box 31700
Omaha, Nebraska 68131
(402) 558-0600

Children's Liver Foundation, Inc.
(CLF)
78 South Orange Ave., Suite 202
South Orange, NJ 07079
(201) 761-1111

Gluten Intolerance Group of North America
(GIG)
P.O. Box 23055
Seattle, WA 98102
(206) 325-6980

National Foundation for Ileitis and Colitis, Inc.
(NFIC)
44 Park Avenue
New York, NY 10016-7374
(212) 685-3440

Hematologic Groups

Cooley's Anemia Foundation, Inc.
105 East 22nd St., Suite 911
New York, NY 10010
(212) 598-0911

Fanconi Anemia Support Group
66 Club Road, Suite 390
Eugene, OR 97401
(503) 687-4658

Hemochromatosis Research Foundation, Inc.
(HRF)
P.O. Box 8569
Albany, NY 12208
(518) 489-0972

Hereditary Hemorrhagic Telangfectasia (HHT) Foundation, Inc.
P.O. Box 8087
New Haven, CT 06530
(313) 561-2537

Histiocytosis-X Association of America, Inc.
609 New York Road
Glassboro, NJ 08028
(609) 881-4911

Iron Overload Diseases Association, Inc.
(IOD)
224 Datura Street, Suite 311
West Palm Beach, FL 33401
(305) 659-5616,5677

National Association for Sickle Cell Disease, Inc.
(NASCD)
3345 Wilshire Boulevard, Suite 1106
Los Angeles, CA 90010
(800) 421-8453, (213) 736-5455

National Hemophilia Foundation
(HANDI)
The Soho Building
110 Greene Street, Room 406
New York, NY 10012
(212) 431-8541

Thrombocytopenia Abset Radius Syndrome Association
(TARSA)
212 Sherwood Drive RD 1
Linwood, NJ 08221
(609) 927-0418

Immunologic Groups

American Lupus Society
23751 Madison Street
Torrance, CA 90505
(213) 373-1335

Immune Deficiency Foundation
(IDF)
Courthouse Square
3565 Ellicott Mills Drive, Unit B2
Ellicott City, MD 21043
(410) 461-3127

Lupus Foundation of America, Inc.
(LFA)
4 Research Place, Suite 180
Rockville, MD
(301) 670-9292

Sjogren's Syndrome Foundation, Inc.
(SSF)
382 Main Street
Port Washington, NY 11050
(516) 767-2866

Kidney Groups

American Association of Kidney Patients
111 South Parker Street #403
Tampa, Florida 33607
(800) 749-2257

Polycystic Kidney Research Foundation
(PKR)
922 Walnut
Kansas City, MO 64106
(816) 421-1869

Mental Health Group

Depression and Related Affective Disorders Association, Inc.
(DRADA)
Johns Hopkins Hospital
Meyer 4–181
601 North Wolfe Street
Baltimore, MD 21205
(301) 955-4647

Volunteer Organizations

Metabolic Groups

American Diabetes Association
1660 Duke Street
Alexandria, VA 22314
(703) 549-1500

American Porphyria Foundation
P.O. Box 22712
Houston, TX 77227
(713) 266-9617

Association for Glycogen Storage Disease
Box 896
Durant, IA 52747
(319) 785-6038

Association of Neuro-Metabolic Disorders
5223 Brookfield Lane
Sylvania, OH 43560
(419) 885-1809

Cystic Fibrosis (CF) Foundation
6931 Arlington Road
Bethesda, MD 20814
(800) FIGHT CF; (301) 951-4422

Cystinosis Foundation, Inc.
17 Lake Avenue
Piedmont, CA 94611
(510) 889-6644

Dysautonomia Foundation, Inc.
370 Lexington Ave.
New York, NY 10017
(212) 889-5222

Foundation for the Study of Wilson's Disease, Inc.
5447 Palisade Avenue
Bronx, NY 10471
(212) 430-2091

Juvenile Diabetes Foundation (JDF) International
432 Park Avenue South, 16th Floor
New York, NY 10016
(212) 889-7575

Lowe's Syndrome Association, Inc.
222 Lincoln St.
West Lafayette, IN 47906
(317) 743-3634

Malignant Hyperthermia Association of the United States
(MHAUS)
P.O. Box 191
Westport, CT 06881
(203) 655-3007

Maple Syrup Urine Disease Family Support Group
(MSUD)
RR #2, Box 24-A
Flemingsburg, KY 41041
(606) 849-4679

ML (Mucolipidosis) IV Foundation
6 Concord Drive
Monsey, NY 10952
(919) 425-0639

National Gaucher Foundation, Inc.
(NGF)
19241 Montgomery Village Ave., Suite E-21
Gaitherburg, MD 20879
(301) 990-3800

National Mucopolysaccharidoses (MPS) Society, Inc.
17 Kraemer Street
Hicksville, NY 11801
(516) 931-6338

National Organization for Albinism and Hypopigmentation
(NOAH)
1500 Locust Street, Suite 1811
Philadelphia, PA 19102
(215) 545-2322

National Tay-Sachs and Allied Diseases Association, Inc.
(NTSAD)
2001 Beacon Street
Brookline, MA 02146
(617) 277-4463

Organic Acidemia Association, Inc.
522 Lander Street
Reno, Nevada 89509
(702) 322-5542

United Leukodystrophy Foundation, Inc.
(ULF) (Canavan disease)
2304 Highland Drive
Sycamore, IL 60178
(815) 895-3211; (800) 728-5483

Williams Syndrome Association
(WSA)
P.O. Box 5297
Ballwin, Missouri 63022
(314) 227-4411

Wilson's Disease Association
P.O. Box 75324
Washington, DC 20013
(703) 636-3003

Zain Hansen M.P.S. (Mucopolysacharridoses) Foundation
P.O. Box 4786
1200 Fernwood Drive
Arcata, CA 95521
(707) 822-5241

Musculoskeletal Groups

Arthritis Foundation/American Juvenile Arthritis, Organization
(AJAO)
1314 Spring Street NW
Atlanta, GA 30309
(404) 872-7100

Freeman-Sheldon Parent Support Group
509 East Northmont Way
Salt Lake City, Utah 87103
(801) 364-7060

National Scoliosis Foundation, Inc.
72 Mt. Auburn St.
Watertown, MA 02172
(617) 926-0397

Osteogenesis Imperfecta Foundation, Inc.
(OIF)
5005 West Laurel Street, Suite 210
Tampa, FL 33607
(813) 282-1161

Osteogenesis Imperfecta National Capital Area, Inc.
Box 941
1311 Delaware Avenue SW
Washington, DC 20024
(202) 265-1614

Paget's Disease Foundation, Inc.
(PDF)
165 Cadman Plaza East
Brooklyn, NY 11201
(718) 596-1043

Scoliosis Association, Inc.
P.O. Box 51353
Raleigh, NC 27609
(919) 846-2639

Neurologic Groups

Acoustic Neuroma Association
3109 Maple Drive, N.E. #406
Atlanta, GA 30305
(404) 237-8023

Alzheimer's Disease and Related Disorders Association, Inc.
(ADRA)
900 N. Michigan Avenue, Suite 1000
Chicago, IL 60611
(312) 335-8700

American Narcolepsy Association
(A.N.A.)
P.O. Box 26230
San Francisco, CA 94126
(415) 788-4793

American Parkinson Disease Association
(APDA)
116 John Street, Suite 417
New York, NY 10028
(800) 223-2732; (212) 732-9550

Batten's Disease Support and Research Association
2600 Parsons Avenue
Columbus, Ohio 43207
(800) 448-4570

Epilepsy Foundation of America
(EFA)
4351 Garden City Drive
Landover, MD 20785
(301) 459-3700

Friedreich's Ataxia Group in America, Inc.
(FAGA)
P.O. Box 11116
Oakland, CA 94611
(510) 655-0833

Hereditary Disease Foundation
1427 7th Street #2
Santa Monica, CA 90401
(310) 458-4138

Huntington's Disease Society of America, Inc.
(HDSA)
140 West 22nd Street
New York, NY 10011-2420
(212) 242-1968

International Joseph Disease Foundation, Inc.
(IJDF)
P.O. Box 2550
Livermore, CA 94551
(510) 443-4600

International Rett Syndrome Association, Inc.
(IRSA)
9121 Piscataway Road, Suite 2B
Clinton, MD 20735
(301) 856-3334

National Hydrocephalus Foundation
(NHF)
400 North Michigan Avenue, Suite 1102
Chicago, IL 60611
(815) 467-6548

National Neurofibromatosis Foundation, Inc.
141 Fifth Avenue, Suite 7S
New York, NY 10010
(800) 323-7983; (212) 460-8980

National Parkinson Foundation, Inc.
(NPF)
1501 Northwest Ninth Avenue
Bob Hope Road
Miami, FL 33136-1491
(800) 327-4545 (outside FL); (305) 547-6666

National Spasmodic Torticollis Association
P.O. Box 873
Royal Oak, MI 48068-0873
(313) 775-1367; (313) 547-2189

National Tuberous Sclerosis Association, Inc.
(NTSA)
8000 Corporate Drive, Suite 100
Landover, MD 20785
(800) CAL-NTSA; (301) 459-9888

Neurofibromatosis, Inc.
8855 Anapolis Rd., Suite 110
Lanham, MD 20706
(301) 577-8984
(800) 442-6825

Parkinson's Disease Foundation, Inc.
(PDF)
650 West 168th Street
New York, NY 10032-9982
(800) 457-6676; (212) 923-4700

Reflex Sympathetic Dystrophy Syndrome Association
(RSDSA)
822 Wayside Lane
Haddonfield, NJ 08033
(609) 428-6510; (609) 428-6980

Spina Bifida Association of America
(SBAA)
4590 MacArthur Blvd., N.W., Suite 250
Washington, DC 20007
(202) 944-3285

Sturge-Weber Foundation
P.O. Box 460931
Aurora, CO 80046
(303) 360-7290

Tourette Syndrome Association, Inc.
(TSA)
42–40 Bell Boulevard
Bayside, NY 11361
(718) 224-2999

Tuberous Sclerosis Association of America, Inc.
(TSAA)
P.O. Box 1305
Middleboro, MA 02370
(617) 947-8893

Neuromuscular Groups

Amyotrophic Lateral Sclerosis Association, Inc.
(ALSA)
21021 Ventura Boulevard, Suite 321
Woodland Hills, CA 91364
(818) 340-7500

AVENUES—National Support Group for Arthrogryposis Multiplex Congenita
P.O. Box 5192
Sonora, CA 95370
(209) 928-3688

Benign Essential Blepharospasm Research Foundation, Inc.
(BEBRF)
P.O. Box 12468
Beaumont, TX 77706
(409) 832-0788

CMT (Charcot-Marie-Tooth) Association
601 Upland Avenue
Crozer Mills Enterprise Center
Upland, PA 19015
(215) 499-7486

Dystonia Medical Research Foundation
(DMRF)
One East Wacker Dr., Suite 2900
Chicago, IL 60601
(312) 755-0198

Families of S.M.A. (Spinal Muscular Atrophy)
P.O. Box 1465
Highland Park, IL 60035
(708) 432-5551

Muscular Dystrophy Association
(MDA)
3300 East Sunrise Drive
Tucson, Arizona 85718
(602) 529-2000

Myasthenia Gravis Foundation, Inc.
(MGF)
53 West Jackson #660
Chicago, IL 60604
(312) 427-6252

Myoclonus Families United
1564 East 34th Street
Brooklyn, NY 11234
(718) 252-2133

National Ataxia Foundation
15500 Wayzata Boulevard #750
Wayzata, MN 55391
(612) 473-7666

National Multiple Sclerosis Society
205 East 42nd Street
New York, NY 10017
(212) 986-3240

Short Stature Groups

Human Growth Foundation
(HGF)
7777 Leesburg Pike
Falls Church, Virginia 22043
(703) 883-1773

Little People of America, Inc.
(LPA)
P.O. Box 94897
Washington, DC 20016
(301) 589-0730

Parents of Dwarfed Children
11524 Colt Terrace
Silver Spring, MD 20902
(301) 649-3275

Skin Groups

Dystrophic Epidermolysis Bullosa Research Association of America, Inc.
(D.E.B.R.A.)
141 Fifth Avenue, Suite 7S
New York, NY 10010
(212) 995-2220

Foundation for Ichthyosis and Related Skin Types, Inc.
(F.I.R.S.T.)
P.O. Box 20921
Raleigh, NC 27619
(919) 782-5728

National Congenital Port Wine Stain Foundation
125 East 63rd Street
New York, NY 10021
(212) 755-3820

National Foundation for Ectodermal Dysplasias
(NFED)
219 East Main Street
Mascoutah, IL 62257
(618) 566-2020

United Scleroderma Foundation, Inc.
(USF)
P.O. Box 399
Watsonville, CA 95077
(408) 728-2202

Xeroderma Pigmentosum Registry
UMDNJ, New Jersey Medical School
Dept. of Pathology, Room C-520
Medical Science Building
100 Bergen Street
Newark, NJ 07103
(201) 456-6255

Visual Groups

American Foundation for the Blind, Inc.
(AFB)
15 West 16th Street
New York, NY 10011
(212) 620-2000

Association for Macular Diseases, Inc.
210 East 64th Street
New York, NY 10021
(212) 605-3719

Blind Children's Fund
230 Central Street
Auburndale, MA 02166-2399
(617) 332-4014

National Association for Parents of the Visually Impaired, Inc.
(NAPVI)
2180 Linway Drive
Beloit, Wisconsin 53511
(608) 362-4945

National Association for Visually Handicapped
22 West 21st Street, 6th Floor
New York, NY 10010
(212) 889-3141

National Society to Prevent Blindness
500 East Remington Road
Shaumberg, IL 60173
(312) 843-2020

Parents and Cataract Kids (PACK)
179 Hunters Lane
Devon, PA 19333
(215) 293-1917; (215) 721-9131; (215) 352-0719

RP Foundation Fighting Blindness
1401 Mt. Royal Avenue
Baltimore, MD 21217
(410) 225-9400

Glossary

Adenine   One of the bases that forms a part of DNA or RNA. It is abbreviated by the letter *A*.

Agar   A gelatinlike substance obtained from seaweed. When used in petri dishes along with a nutrient broth as food, agar enables microbiologists to grow bacterial colonies.

Agar plate   A Petri dish containing solid agar.

Amino acid   A small molecule that serves as a subunit of protein. The twenty different amino acids have a common structure shown below. The letter *R* represents chemical side chains, which are different for each amino acid. The chemical properties of the side chains help determine how a protein folds; thus the arrangement of amino acids dictates the three-dimensional structure of a protein.

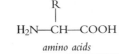

amino acids

alanine (ala)	leucine (leu)
arginine (arg)	lysine (lys)
asparagine (asn)	methionine (met)
aspartic acid (asp)	phenylalanine (phe)
cysteine (cys)	proline (pro)
glutamine (gln)	serine (ser)
glutamic acid (glu)	threonine (thr)
glycine (gly)	tryptophan (trp)
histidine (his)	tryosine (tyr)
isoleucine (ile)	valine (val)

Aminoacyl-tRNA synthetases Members of a class of enzyme that link specific amino acids with specific transfer RNA molecules. One synthetase recognizes one particular type of transfer RNA and one particular type of amino acid.

Amniocentesis A surgical procedure in which a hollow needle is inserted through the abdominal wall and into the uterus of a pregnant woman to obtain a sample of amnionic fluid from around a developing fetus. The cells in the fluid are then analyzed for genetic disease.

Antibiotic A substance produced by a microorganism that inhibits the growth of bacteria, often killing them. Most antibiotics in clinical use have been modified extensively to be more potent. Common examples are streptomycin, erythromycin, penicillin, ampicillin, and tetracycline.

Antibiotic-resistance gene A gene that codes for a protein that allows a bacterium to live in the presence of a drug that normally would kill it. Some resistance genes change the target of the drug so it no longer binds the drug. Others cause active secretion of the drug, and still others break down the drug. Plasmids often contain such genes.

Antibodies Proteins that recognize and bind to foreign proteins such as those found on the surfaces of viruses and the surfaces of bacteria. After binding the foreign protein (often called an antigen), the antibody can participate in a variety of reactions that lead to the destruction of the antigen. Antibodies are an important component of the immune system, which serves to guard us from attack by microorganisms.

Anticodon A particular three-nucleotide region in transfer RNA that is complementary to a specific three-nucleotide codon in messenger RNA. Alignment of codons and anticodons is the basis for establishing the order of amino acids in a protein chain.

Antigen A chemical, protein, or microorganism that is recognized by, and attaches to, an antibody.

Antisense RNA An RNA molecule that is the complement of another RNA molecule. The two RNA molecules have the ability to form a hybrid. Antisense RNA molecules can be designed to hybridize with particular mRNA molecules and thereby prevent the mRNA from being translated.

Assay   A method or way of measuring chemical compounds.

Atom   A particle composed of a nucleus (protons and neutrons) and electrons. Common atoms are carbon, oxygen, nitrogen, and hydrogen. These atoms differ from one another by having different numbers of protons, neutrons, and electrons. Groups of atoms bonded together produce molecules.

ATP   Adenosine triphosphate, a relatively small molecule that serves as an energy carrier and as one of the precursors to RNA. ATP has high-energy bonds that are easily broken by enzymes to release the energy needed to drive some cellular chemical reactions.

B lymphocyte   A type of cell in mammals that produces antibodies.

Bacterial culture   A batch of bacterial cells grown either on solid agar or in a brothlike solution.

Bacteriophage   A virus that attacks bacteria; also called a phage.

Bacterium (plural bacteria)   A one-celled organism lacking a nucleus, mitochondria, and chloroplasts. Although many biochemical properties of bacteria differ from those of higher organisms, the basic features of chemical reactions are very similar in bacteria and in human beings.

Base   (1) A flat, ring structure, containing nitrogen, carbon, oxygen, and hydrogen, that forms part of one of the nucleotide links of a nucleic acid chain. The different bases are adenine, thymine, guanine, cytosine, and uracil, commonly abbreviated A,T,G,C, and U. (2) A hydrogen ion acceptor, such as sodium hydroxide (lye).

Base pair (bp)   Two bases, one in each strand of a double-stranded nucleic acid molecule, which are opposite each other. The bases of a base pair are attracted to each other by weak chemical interactions. Only certain pairs form: A•T, G•C, and A•U.

Broth   A liquid culture medium used to grow bacteria. One common type contains yeast extract, beef extract, table salt, and water.

Carcinogen   A chemical that causes cancer, generally by altering the structure of DNA (*See* mutagen).

Cell   The smallest unit of living matter capable of self-perpetuation; an organized set of chemical reactions capable of reproduction. A cell is bounded by a membrane that separates the inside of the cell from the outer environment. Cells contain DNA (where information is stored), ribosomes (where proteins are made), and mechanisms for converting energy from one form to another.

Cell extract or lysate   A mixture of cellular components obtained by mechanically or enzymatically breaking cells. The cell extract is the starting material from which biochemists obtain enzymes, RNA, and DNA.

Cell wall   A thick, rigid structure surrounding cells of certain types, especially bacterial and plant cells. Cell walls are often composed of complex sugars.

Centrifuge   A machine that uses centrifugal force generated by a spinning motion to separate molecules of various sizes and densities. Centrifuges can create forces hundreds of thousands of times that of gravity, making it possible to quickly separate molecules on the basis of size and shape. Merry-go-rounds and the spin cycle mechanisms of automatic clothes-washing machines are examples of centrifuges.

Centrifuge rotor   The part of a centrifuge that holds test tubes and spins at high speed.

Chemical reaction   A rearrangement of atoms to produce a set of molecules that are different from the starting set of molecules.

Chromosome   A subcellular structure containing a long, discrete piece of DNA plus the proteins that organize and compact the DNA.

Clone   (1) Noun—a group of identical cells, all derived from a single ancestor. (2) Verb—to perform or undergo the process of creating a group of identical cells or identical DNA molecules derived from a single ancestor.

Cloning vehicles   Small plasmid, phage, or animal virus DNA molecules into which a DNA fragment is inserted so the fragment can be transferred from a test tube into a living cell. Cloning vehicles are capable of multiplying inside living cells. Thus, if a cloning vehicle transfers a specific fragment of DNA into a cell that is also multiplying, all the progeny of that cell will contain identical copies of the vehicle and the transferred DNA fragment.

Code (genetic)   The system in which the arrangement of nucleotides in DNA represents the arrangement of amino acids in protein.

Codon   Three nucleotides in DNA and RNA whose precise order corresponds to one of the twenty amino acids. In addition, special codons that do not code for any amino acid act as stop signals during protein synthesis. In some cases, several different codons encode the same amino acid.

Colony A visible cluster of bacteria on a solid surface. All members of the colony arise from a single parental cell, and the colony is considered to be a clone. All members are identical. A colony generally contains millions of individual cells.

Complementary Describing two objects having shapes that allow them to fit together very closely: plugs and sockets, locks and keys, As and Ts or Us, Gs and Cs.

Complementary base-pairing rule Only certain nucleotides can align opposite each other in the two strands of DNA: G pairs with C; A pairs with T (or U in RNA).

Complementary DNA (cDNA) DNA synthesized from RNA in test tubes using an enzyme called reverse transcriptase. The DNA sequence is thus complementary to that of the RNA. Complementary DNA is usually made with radioactive nucleotides and is used as a hybridization probe to detect specific RNA or DNA molecules.

Cystic fibrosis A recessive genetic disease caused by the inability of cells to properly secrete salt. Among the symptoms is formation of thick mucus in the lungs.

Cytology The study of cell structure, often using microscopy. In the present context, a commercial cytology laboratory examines the structure of chromosomes for genetic disorders.

Cytosine One of the bases that forms a part of DNA or RNA. It is usually abbreviated with the letter *C*.

Denature To unfold, to become inactive. In reference to DNA, denaturation means conversion of double-stranded DNA into single-stranded DNA. In reference to proteins, denaturation means unfolding of the protein.

Deoxyribonucleic acid *See* DNA.

Dissolve To disperse a solid substance in a liquid.

DNA Deoxyribonucleic acid. DNA is a long, thin, chainlike molecule that is usually found as two complementary chains and is often hundreds to thousands of times longer than the cell in which it resides (it is tightly wrapped to fit inside). The links or subunits of DNA are the four nucleotides called adenylate, cytidylate, thymidylate, and guanylate. The precise arrangement of these four subunits is used to store all information necessary for life processes.

DNA ligase An enzyme that joins two separate DNA molecules together end to end.

DNA polymerase An enzyme complex that makes new DNA using the information contained in old DNA.

DNA replication The process of making DNA. DNA is always made from preexisting information in DNA (or, in special cases, from RNA). DNA replication involves a number of different enzymes.

E. coli *Escherichia coli,* a type of bacterium are commonly found in the digestive tracts of many mammals, including humans.

Egg (1) Germ cell produced by a female; also called an ovum. (2) An animal embryo, along with a food supply, enclosed by a shell or membrane.

Electron micrograph A photograph taken using an electron microscope.

Electron microscope An instrument that is similar to a light microscope but uses a beam of electrons to expose the film rather than a beam of light. Because the effective wavelength of electrons is much shorter than that of light, objects that are measured in millionths of a centimeter can be seen using an electron microscope.

Electrophoresis A process in which an electric field is used to separate molecules of different charge.

Element One of slightly more than 100 distinct types of matter, which singly or in combination compose all materials of the universe. An atom is the smallest representative unit of an element.

Embryo A plant or an animal in an early stage of development, generally still contained in a seed, egg, or uterus.

Encode Contain a nucleotide sequence specifying that one or more specific amino acids be incorporated into a protein.

Endonuclease An enzyme that cuts DNA or RNA at points inside the molecule (i.e., away from the ends).

Envelope A covering or coat; the outermost coat of an animal virus.

Enzyme A protein molecule, or occasionally an RNA molecule, specialized to catalyze (accelerate) a biological chemical reaction. Generally enzyme names end in -*ase*.

Equilibrium   The absence of *net* movement one way or another.

Expression   *See* gene expression.

Expression vector   A plasmid or virus in which the DNA contains an active but easily controlled promoter, downstream from which a gene of interest can be inserted. Following induction of the promoter, the protein of interest can be produced in large amounts, sometimes comprising up to 40 percent of the total cellular protein of the bacterium that carries the vector.

Extract   (1) Verb—to separate one type of molecule from all others, to purify. (2) Noun—a mixture of molecules obtained by breaking cells.

Fetus   (adj. fetal) An embryo in a late stage of development, but still in the uterus.

Fission   A type of cell division in which a parental cell divides in half to form two daughter cells.

Five-prime (5′) and three-prime (3′) ends   The backbone of a nucleic acid molecule is composed of repeating phosphate and sugar subunits such that on one side of the sugar the phosphate is linked to the 5′ carbon of the sugar and on the other side the phosphate is linked to the 3′ carbon of the sugar. When a chain is broken, the break generally occurs between the phosphate and the sugar. This produces two different ends. If only the sugar is considered, a 5′ carbon will be at one end (the 5′ end) and a 3′ carbon will be at the other (the 3′ end). These terminal carbons generally have a phosphate or an -OH group attached to them.

Gel electrophoresis   A method for separating molecules based on their size and electric charge. Molecules are driven through a gel (e.g., gelatin) by being placed in an electric field. The speed at which they move depends on their size and charge.

Gene   A small section of DNA that contains information for construction of one protein molecule or in special cases for construction of transfer RNA or ribosomal RNA.

Gene cloning   A way to use microorganisms to produce millions of identical copies of a specific region of DNA.

Gene expression   The process of making the product of a gene. This involves transferring information, via messenger RNA, from a gene to ribosomes, where a specific protein is made.

Genetic engineering The manipulation of the information content of an organism to alter the characteristics of that organism. Genetic engineering may use simple methods like selective breeding or complicated ones like gene cloning.

Genome The primary repository of genetic information for an organism; generally refers to the total set of DNA molecules from each of the chromosomes. In certain viruses, genetic information is stored in an RNA form.

Germ cells A particular type of cell (sperm and eggs) responsible for creating the next generation; also called gametes. In most higher organisms, body cells contain two sets of chromosomes; germ cells contain only one set. Thus when two germ cells join together, the resulting cell (zygote) has two sets of chromosomes. This cell then produces new body (somatic) cells.

Guanine One of the bases that forms a part of DNA or RNA. It is abbreviated with the letter G.

Homologous Corresponding or similar in position; describing regions of DNA molecules that have the same nucleotide sequence. Since DNA has two complementary strands, complementary base-pairing can occur between homologous regions in two different DNA molecules. Homologous also refers to regions of DNA, RNA, or protein that are similar because of a common ancestry.

Host An organism that provides the life-support system for another organism, virus, or plasmid. *E. coli* is a host for certain plasmids that exist inside the bacterium, and we are the host for *E. coli,* for these bacteria live inside us.

Hybrid, nucleic acid A double-stranded nucleic acid in which the two complementary strands differ in origin. One strand can be RNA, and the other DNA or both strands can be either RNA or DNA.

Induce Cause to happen, often with reference to gene expression. Specific molecules called inducers bind to certain repressors and prevent the repressor from binding to DNA. That allows gene expression to occur.

Infectious Capable of invading a host.

Insulin A protein involved in the control of sugar metabolism in mammals. Insulin is made by cells of the pancreas.

Kinase An enzyme that adds a phosphate to another molecule.

Leader A region of film, RNA, or protein that precedes the region of primary information content. In RNA, the leader extends from the first nucleotide at the 5′ end to the codon specifying the first amino acid in the protein. A protein leader, if present, is usually defined as a region at the amino terminal end that is cut from the protein during movement of the protein across membranes.

Lysozyme An enzyme that breaks down bacterial cell walls. Lysozyme can be obtained from egg white or tears.

Macromolecule A very large molecule, usually composed of many smaller molecules joined together.

Messenger RNA (mRNA) RNA used to transmit information from a gene in DNA to a ribosome, where the information is used to make protein.

Metabolism A collective term for all the chemical reactions involved in a particular aspect of life. For example, sugar metabolism includes the reactions that occur in the body during the production, use, and breakdown of sugars.

Micrometer One millionth of a meter (1 meter = about 39 inches).

Milligram One thousandth of a gram (28 grams = 1 ounce).

Millimeter One thousandth of a meter (1 meter = about 39 inches).

Mitochondrion A specialized intracellular structure that converts chemical energy from one form to another. Mitochondria contain DNA molecules that encode some mitochondrial proteins.

Molecule A group of atoms tightly joined together. The arrangement of atoms is very specific for a given molecule, and this arrangement gives each molecule specific chemical and physical properties. The oxygen molecule we breathe is two oxygen atoms bonded together. Paper is largely cellulose molecules, which are giant molecules containing carbon, oxygen, and hydrogen.

Monolayer A layer of cells one cell thick.

Mutagen An agent that increases the rate of mutation by causing changes in the nucleotide sequences of DNA (*see* carcinogen).

Mutant   An organism whose DNA has been changed relative to the DNA of the dominant members of the population.

Mutations   Errors in DNA, often arising during DNA replication, that cause incorrect amino acids to be inserted into proteins.

Neonatal   Newborn.

Nuclease   A general term for an enzyme that cuts DNA or RNA.

Nucleic acid   DNA, RNA, or a DNA•RNA hybrid.

Nucleic acid hybridization   A process in which two single-stranded nucleic acids are allowed to base pair and form a double helix. The process makes it possible to use one nucleic acid to detect the presence of another having nucleotide sequence similarity.

Nucleotide   One of the building blocks of nucleic acids. A nucleotide is composed of three parts: A base, a sugar, and a phosphate. The sugar and the phosphate form the backbone of the nucleic acid, while the bases lie perpendicular to the backbone and flat like steps of a spiral staircase. DNA is composed of deoxyadenylate, deoxythymidylate, deoxyguanylate, and deoxycytidylate, four different kinds of nucleotide represented by the letters A, T, G, and C. *See also* sequence.

Nucleotide pair   Two nucleotides, one in each strand of a double-stranded nucleic acid molecule, that are attracted to each other by weak chemical interactions between the bases. Only certain pairs form: A•T, G•C, and A•U.

Nucleus   (1) The core of an atom consisting of protons and neutrons; (2) a distinct subcellular structure containing chromosomes.

Oligonucleotide   A short piece of DNA or RNA containing three or more nucleotides. The oligonucleotides used in gene cloning are generally less than 100 nucleotides long, but in formal terms an oligonucleotide can be much longer.

Operator   A region on DNA capable of interacting with a repressor, thereby controlling the functioning of an adjacent gene.

Operon   A series of genes transcribed into a single RNA molecule. Operons allow coordinated control of a number of genes whose products have related functions.

Organism   One or more cells organized in such a way that the unit is capable of reproduction.

Origin of replication A special nucleotide sequence that serves as a start signal for DNA replication.

Pathogen A disease-causing agent (e.g., viruses that cause polio, mumps, and measles; bacteria that cause cholera and leprosy).

Penicillin An antibiotic that kills *E. coli* and many other bacteria by blocking formation of new cell walls. Penicillin is produced by a mold.

Peptide A short chain of amino acids; a fragment of a protein.

Peptide bond The type of chemical bond that links two adjacent amino acids together in a protein chain.

Phage A virus that attacks bacteria; abbreviation for bacteriophage.

Phage plaques Clear zones, created by bacteriophages killing bacteria, in a lawn of bacteria on an agar plate.

Phosphate A chemical unit in which four oxygen atoms are joined to one phosphorus atom. The backbones of DNA and RNA are alternating phosphate and sugar units.

Plasmids Small, circular DNA molecules found inside bacterial cells. Plasmids reproduce every time the bacterial cell reproduces.

Polymerase chain reaction A test tube reaction in which a specific region of DNA is amplified many times by repeated synthesis of DNA using DNA polymerase and specific primers to define the ends of the amplified region.

Polyprotein A long protein that is cleaved into several smaller proteins. The smaller proteins are thought to be the functional forms.

Precipitate Molecules that are clumped together so that they fail to pass through a filter. Precipitates are large aggregates and settle out of solution rapidly, much like silt out of river water.

Prenatal Before birth.

Primer A piece of DNA or RNA that provides an end to which DNA polymerase can add nucleotides.

Probe A DNA or RNA molecule, usually radioactive, that is used to locate a complementary RNA or DNA by hybridizing to it. Often a probe is used to identify bacterial colonies that contain cloned genes and to detect specific nucleic acids following separation by gel electrophoresis.

Product The new molecules produced by a chemical reaction.

Progeny Offspring.

Promoter A short nucleotide sequence in DNA where RNA polymerase binds and begins transcription.

Protease An enzymatic protein that breaks down other proteins.

Protein A class of long, chainlike molecules often containing hundreds of links called amino acids. Twenty different amino acids are used to make proteins. The thousands of different proteins serve many functions in the cell. As enzymes, they control the rate of chemical reactions, and as structural elements they provide the cell with its shape. Members of this group of molecules are also involved in cell movement and in the formation of cell walls, membranes, and protective shells. Some proteins also help package the long DNA molecules into chromosomes.

Protein synthesis The process of linking amino acids to form protein. *See* translation.

Purify To separate or isolate one type of molecule away from other types.

Radioactive The state in which a substance (a molecule in the context of this book) contains an unstable atom that spontaneously emits a high energy particle or radiation. The emission is detectable by photographic film, by Geiger counter, and by other instruments. Gene cloners generally use radioactive hydrogen, carbon, or phosphorus, each of which is commercially available. Radioactive uranium and plutonium are used in nuclear reactors.

Radioactive tracer A radioactive atom that is incorporated into a specific molecule such that the molecule can be detected and measured by the presence of the radioactivity. It is much easier to measure small amounts of radioactivity than small amounts of particular chemicals.

Recognition site A short series of nucleotides specifically recognized by a protein, usually leading to the binding of that protein to the DNA at or near the point of the recognition sequence. Once the protein has bound to the DNA, it may cut, modify, or cover the DNA, depending on the function of the protein.

Recombinant DNA molecule A DNA molecule containing two or more regions of different origin (e.g., plasmid DNA joined to a fragment of human DNA).

Recombination The breaking and rejoining of DNA strands to produce new combinations of DNA molecules. Recombination is a natural process that generates genetic diversity. Specific proteins are involved in the recombination process.

Replication fork   The point at which the two parental DNA strands separate during DNA replication.

Repression   A process in which a protein molecule binds to a specific region of DNA and prevents specific gene expression.

Repressor   A protein molecule that is capable of preventing transcription of a gene by binding to DNA in or near the gene.

Restriction endonucleases   Enzymes that cut DNA at specific nucleotide sequences. The intracellular function of this class of enzyme is to protect cells against invasion by foreign DNA. Biologists use these enzymes as scissors to cut DNA in specific places.

Restriction fragment length polymorphism (RFLP)   A region of DNA that has several forms (the region differs from one individual to another) such that cleavage of DNA using a restriction endonuclease generates a DNA fragment whose size varies from one individual to another.

Restriction mapping   A procedure that uses restriction endonucleases to produce specific cuts in DNA. The positions of the cuts can be determined relative to each other to form a crude map.

Retrovirus   A type of animal virus having a life cycle that involves conversion of genetic information from an RNA form to a DNA form.

Reverse transcriptase   An enzyme purified from retroviruses that makes DNA from RNA.

RFLP   *See* restriction fragment length polymorphism.

Ribonucleic acid   *See* RNA.

Ribosomes   Large, ball-like structures that act as workbenches where proteins are made. A bacterial ribosome consists of two balls, a small one called 30S and a larger one called 50S (30S and 50S refer to the speed at which the particles sediment during centrifugation). Ribosomes are composed of special RNA molecules (ribosomal RNA) and about 50 specific proteins (ribosomal proteins).

Ribozyme   An RNA molecule that acts catalytically to cleave itself or another RNA molecule. Some ribozymes have the ability to join two RNA molecules together end-to-end.

RNA   Long, thin, chainlike molecules in which the links or subunits are the four nucleotides called adenylate, cytidylate, uridylate, and guanylate (abbreviated with the letters A, C, U, and G). RNA molecules are used to transfer, and sometimes store, genetic information. Some RNA molecules serve as structural parts of cellular components (ribosomes), and some (transfer RNA) help connect amino acids in the correct order when proteins are being made. Some RNA molecules (ribozymes) have enzymatic activity and serve as catalysts to accelerate specific chemical reactions.

RNA•DNA hybrid   A double-stranded molecule in which one strand is RNA and one is DNA. The nucleotide sequences in the DNA and RNA strands are complementary.

RNA polymerase   The enzyme complex responsible for making RNA from DNA. RNA polymerase binds at specific nucleotide sequences (promoters) in front of genes in DNA. It then moves through a gene and makes an RNA molecule that contains the information contained in the gene. Bacterial RNA polymerase makes RNA at a rate of about 65 nucleotides per second.

RNA splicing   The process of removing regions from RNA. The removed regions are called introns, and the regions spliced together are called exons.

RNA tumor virus   A type of RNA-containing virus that produces tumors in animals or converts normal cells in culture into tumor cells.

Sequence   "The order of." In reference to DNA or RNA, sequence means the order of nucleotides.

Somatic   Pertaining to the body. When referring to a type of cell, somatic means body cell rather than a sperm- or egg-producing cell.

Sonogram   An image of internal body parts obtained by use of high frequency sound. In the present context, sonograms provide an image of a fetus in a pregnant woman.

Southern hybridization (Southern blotting)   A method for transferring DNA from an agarose or acrylamide gel to nitrocellulose paper or nylon membrane followed by hybridization to a radioactive probe.

Sperm   Germ cell produced by a male.

Sterile Without life; generally referring to an instrument or a solution that has been heated to kill organisms that may have been on or in it. Wire is sterilized by heating in a flame until it is red hot. Culture medium (e.g., broth) is sterilized by heating in a pressure cooker (autoclave). Sterile also means unable to reproduce.

Submicroscopic Not visible when examined with a light microscope.

Substrate The molecules on which an enzyme acts.

Subunit One of the pieces that forms a part of a multicomponent structure, such as a link in a chain or a brick in a wall.

Sugar A class of molecule containing particular combinations of carbon, hydrogen, and oxygen. The sugars in DNA and RNA are five-carbon sugars called deoxyribose and ribose, respectively. Glucose, a major constituent of honey, is a sugar containing six atoms of carbon per molecule.

Sugar metabolism A group of biochemical reactions responsible for the formation of sugars and the conversion of sugars into other compounds.

Tetracycline An antibiotic that kills bacteria by blocking protein synthesis.

Tetramers Four subunits, often identical. Many proteins are composed of separate polypeptide chains that act as subunits, associating as a tetramer to form the active protein.

Thymine One of the bases that forms part of DNA. It is not found in RNA. It is usually abbreviated by the letter *T*.

Topoisomerase An enzyme that breaks and rejoins DNA strands in such a way that it changes the number of times one strand crosses the other. Topoisomerases can tie and untie DNA knots, add and subtract twists in the DNA, and link and unlink DNA circles.

Toxin A substance, often a protein in the context of this book, that causes damage to the cells of an organism.

Transcription The process of converting information in DNA into information in RNA. Transcription involves making an RNA molecule using the nucleotide sequence of DNA as a template. RNA polymerase is the enzyme that executes this conversion of information.

Transfer RNAs (tRNAs) Small RNA molecules (each about eighty nucleotides long) that serve as adapters to position amino acids in the correct order during protein synthesis. The ordering by tRNA occurs through base pairing of a region of each tRNA molecule to messenger RNA.

Transformation The process whereby a bacterial cell takes up free DNA such that information in the free DNA becomes a permanent part of the bacterial cell. Often this means introducing a plasmid into a bacterial cell. With animal cells, transformation can also mean the conversion of a normal cell into a tumor cell.

Translation The process of converting the information in messenger RNA into protein. Also called protein synthesis.

Transposase A protein encoded by a gene in a transposon and required for transposition.

Transposition The process whereby one region of DNA moves to another. Transposition often involves duplication of the region that moves.

Transposon A short section of DNA capable of moving to another DNA molecule or to another region of the same DNA molecule.

Ultraviolet light A type of light that has very high energy and is invisible to human beings; blacklight. Nucleic acids absorb ultraviolet light, and instruments are available that measure the amount of absorption. The amount of absorption depends on the amount of nucleic acid present. Thus, by measuring the amount of absorption, it is possible to measure the amount of nucleic acid present.

Uracil One of the bases that forms part of RNA. It is generally not found in DNA. Uracil is abbreviated with the letter *U*.

Vector *See* expression vector.

Virus particles A class of infectious agent usually composed of DNA or RNA surrounded by a protective protein coat.

VNTR A particular type of restriction fragment length polymorphism used for identifying organisms, especially human beings. The letters represent "variable number of tandem repeats."

Yeast One-celled organism commonly used in brewing and baking. Yeast contains a true nucleus and mitochondria. Many biochemical properties of yeast are more closely related to those found in mammals than to those of bacteria.

Sources for Quotations

Page 20: E.P. Fischer and C. Lipson, *Thinking About Science* (1988), page 112.

Page 24: Max Delbruck, *The Harvey Lectures* (1947) series XLI, page 162.

Page 30: Gunther Stent, *Genetics* (1982) vol. 6, pages 8 and 12.

Page 38: Neville Symonds, *Trends in Biochemical Sciences* (June 1988), page 233.

Page 41: Arthur Kornberg, *For the Love of Enzymes* (1989), page 52.

Page 56: C. Allbutt, *British Medical Journal* (April 6, 1918), page 389.

Page 102: K. Bergalis, *AIDS Alert* (July 1991) vol. 6, page 144.

Page 115: The State of Texas vs. David Hicks, Freestone County, Texas, No. 88-134-CR (1989), page 901.

Index

ADA deficiency, 76–83
AIDS, 52, 78, 83, 86–95, 99–104, 130, 149, 164, 173
AZT, 88, 89, 93
Acer, David, 102, 103
Acoustic neuroma, 209
Adenine, 35
Adenosine deaminase (ADA), 76–78
Adenovirus, 82
Agar, 21, 96, 98, 99, 101, 168
Agent Orange, 179
Aging, 132, 137
Albinism, 207
Aldosterone, 136
Allergic reactions, 145
Alzheimer disease, 138, 141, 148, 209
Amino acid, 43, 44, 48, 49, 51, 52, 165
Amnionic fluid, 7, 11
Amyotrophic lateral sclerosis, 213
Anderson, French, 77
Anderson, Tom, 28
Antibiotic, 55, 96, 98, 100, 101, 106, 145, 148
  resistance, 100, 105, 144, 145, 162, 163
Antibody, 88, 97
Antisense DNA, 174
Antisense RNA, 83, 174, 181
Apolipoprotein, 138

Arthritis, 208
Arthrogryposis, 213
Ataxia, 215
Atoms, 8, 24–26, 29, 31, 33, 34, 36, 40, 45, 52, 158, 161
Autism, 199
Avery, Oswald, 30

Bacteria, 9, 20–23
  associated with cystic fibrosis, 9
  gene cloning with, 51, 129, 131, 169
  gene transfer, 163
  growth as colonies, 96, 101
  growth as lawns, 21, 22
  pathogenic, 92, 96–101, 105, 106, 162, 163
Bacteriophage, 20–31, 40, 60
  counting, 21, 22
  infection of bacteria, 21, 22, 30, 31, 60
  mutants, 26, 46, 47
  plaques, 21, 22, 28
Base, 35, 46, 48, 158–61
  complementary pairs, 35, 42, 50, 159–62, 174, 175
Batten disease, 210
Berg, Paul, 51
Bergalis, Kimberly, 102, 103
Blaese, Michael, 77

Blepharospasm, 213
Blindness, 217, 218
Blood doping, 131
Bone marrow, 77

Canavan disease, 150
Cancer, 20, 42, 54, 56, 82, 83, 137–40,
    194
  associated with defective, p53, 140
  colorectal, 195
  detection, 140
  genes associated with, 139, 140, 141
  leukemia, 195
  ovarian, 195
  polyposis, 195
  predisposition to, 139, 140, 149
Carrier of disease (*see* Disease carrier)
Cell, 8, 26, 48
  bacterial, 20–23, 26
  blood, 77, 82, 92, 130, 131
  germ, 54, 83–85
  liver, 81, 138
  skin, 82
  somatic, 54, 83, 85, 183
  stem, 83
Cerebral palsy, 200
Charcot-Marie-Tooth disease, 214
Chemical reactions, 40, 41, 44
Cholesterol, 81, 137, 138
Chromosome, 5, 9, 62, 77–79, 82, 85,
    163, 164
  abnormalities, 17, 18, 139, 158, 184,
    185
  abnormalities, 5p, 196
  abnormalities, 9p, 197
  bacterial, 32
  condensed, 184
  X and Y, 182
  X inactivation, 183

Cleft palate, 198
Clone, 129, 131, 162, 169
Clotting factor, 81, 130
Codon, 45, 46–49, 51, 162, 165
Colitis, 201
Complementary base-pairing, 35, 42,
    50, 160–62, 174, 175
Contraceptives, 146
Cooley's anemia, 202
Cornelia de Lange syndrome, 199
Coronary artery disease, 137, 138
Cowpox virus, 146, 147
Creutzfeldt-Jacob disease, 129, 130
Crick, Francis, 1, 33–35, 45–47
Cystic fibrosis, 8–13, 82, 85, 148, 150,
    181, 205
Cystinosis, 205
Cytosine, 35

DNA, 9, 13
  bacterial chromosomal, 32
  cystic fibrosis test, 13
  evidence as genetic material, 30, 31
  fingerprinting, 108–25, 162
  hybrid with RNA, 160
  information in, 45–51, 54
  isolation and storage, 185, 186
  length, 33
  ligase, 168
  polymerase, 42, 171–73
  polymorphisms (*see* RFLP, VNTR),
    60–64, 109, 116, 183
  recombinant, 129, 132, 133, 169
  replication, 35–37, 40–42, 166, 167,
    171–73
  replication initiation, 42
  structure, 33, 34, 157–62
  typing (*see* DNA fingerprinting)
  viral, 78, 158

Dantrolene, 135
Deafness, 194
Delbrück, Max, 19–31, 36, 38, 40, 47, 49, 51, 151
Dental cavities, 162, 163
Deoxyadenosine, 76
Deoxyadenosine triphosphate, 76
Deoxyinosine, 76
Deoxyribose, 158–60
Depression, 204
Diabetes, 54, 81, 130, 132, 205, 206
Disease carriers
  bacterial, 96–101, 105–107
  hereditary, 9, 10, 13–17, 48, 56–75, 179, 181, 182, 187
  HIV, 87–95, 101–107
Down syndrome, 184, 196, 197
Dwarf, 81, 127–30, 215
Dysautonomia, 205
Dystrophic epidermolysis bullosa, 215

Ectodermal dysplasia, 216
Eggs, 62, 83, 84, 146, 149, 165
Ehlers-Danlos Syndrome, 198
Ellis, Emory, 20, 21, 23
Embryos, 3, 50, 84–86, 149, 165, 183
Enhancer, 165
Enzymes, 40–45, 52, 60, 76–79, 82, 164, 165, 181
Epilepsy, 210
Erythropoietin, 131
Escherichia coli, 23, 40

Familial hypercholesterolemia, 137, 138
Family analysis, 16, 17, 176–86
Fanconi anemia, 202
Fragile X syndrome, 196, 197
Franklin, Rosalind, 33
Friedreich ataxia, 210

Gaucher disease, 150, 206
Gehrig, Lou, 149
Gel electrophoresis, 110, 111
Gene, 5, 9
  anticancer, 139
  cloning, 169
  expression, 48, 49, 165
  inheritance patterns, 5, 10, 58, 176–84
  insertion, 79
  order, 28, 62
  structure, 45
  therapy, 76–86, 138, 150, 175, 181
  therapy, cosmetic, 81
  therapy, germ line, 83
Genes race, 3, 84
Genetic code, 48
Genetic counselling, 72, 189–92
Genetic disclosure, 149, 150
Genetic discrimination, 187, 188
Genetic engineering of food, 143–46
Genetic screening/testing, 3, 14–17, 54, 148
  Canavan disease, 150
  cystic fibrosis, 11, 14, 15, 150
  Gaucher disease, 150
  malignant hyperthermia, 136, 140
  phenylketonuria, 15
  sickle-cell anemia, 15, 71
  Tay-Sachs disease, 14, 150
Germ cells, 54, 83, 84, 150
Germ line gene therapy, 83, 150
Glucocorticoid-remedial aldosteronism, 136
Gluten intolerance, 201
Glycogen storage disease, 205
Growth hormone, 81
Guanine, 35
Gusella, James, 64

HIV (human immunodeficiency virus), 87–95, 101–106, 130, 174
  test for, 88, 92, 93, 103, 173
  transmission, 94–95, 101–104, 106
Hartl, Daniel, 120
Hawking, Stephen, 149
Heart disease, 136, 137, 140
Hematochromatosis, 202
Hemoglobin, 50
Hemophilia, 81, 130, 148, 182, 203
Hemorrhagic telangfectasia, 202
Hershey, Alfred, 23, 31, 49
Histiocytosis, 202
Hormones, 53, 127–33, 145, 146
  bovine, 145, 146
  human growth, 127–32
  human, produced by bacteria, 129, 131, 132, 146
Human genome project, 71, 147
Human growth, 215
Huntington disease, 53, 56–67, 69, 71, 73, 81, 83, 181, 182, 210
Huntington, George, 57
Hydrocephalus, 211
Hydrogen bonds, 158, 161
Hyperactive children, 199

Ichthyosis, 216
Ileitis, 201
Immune deficiency, 203
Immune system, 76, 92, 99–101, 146, 147
Informed consent, 15, 73, 83, 133
Inheritance
  dominant trait, 5, 58, 69, 136, 180–182, 184
  multifactor traits, 183
  patterns, 5, 10, 58, 176–184

recessive trait, 5, 9, 14, 15, 17, 84, 179, 181, 182, 184
  X-linked trait, 179, 182, 184
Insulin, 43, 55, 81, 128, 130
Ion, 8
  chloride, 8
  sodium, 8
Iron overload diseases, 202

Joseph disease, 69, 211

Kidney diseases, 204
Kinase, 76
Kornberg, Arthur, 40–42, 60

Laurence-Moon-Biedl syndrome, 200
Learning disabilities, 199
Leukodystrophy, 207
Lewontin, Richard, 120
Lincoln, Abraham, 68, 149
Liver disorders, 201
Lou Gehrig disease, 132
Lowe syndrome, 206
Lupus, 204
Luria, Salvadore, 23, 30, 31, 49
Lymphocytes, 76, 78, 79, 82, 89

Macrophages, 92
Malignant hyperthermia, 69, 134–36, 141, 206
Maple syrup urine disease, 15, 206
Marfan syndrome, 68, 198
Medical Information Bureau, 73, 75
Mendel, Gregor, 5
Mendel's Laws, 4, 5
Miscarriages, 16–18, 65, 74
Molecular recognition, 52
Molecule, 25–33, 36, 40, 41, 52, 53, 158

Mucolipidosis, 206
Mucopolysaccharidoses, 207
Multiple sclerosis, 215
Muscular dystropy, 214
Mutagen, 168
Mutation, 20, 27, 28, 46–47, 54, 85, 138, 139, 158, 166, 167
  deletion, 166, 167
  frameshift, 47, 166, 167
  missense, 166, 167
  new, 183, 184
  nonsense, 166, 167
  unstable, 66
Myasthenia gravis, 214
Myoglobin, 44

Narcolepsy, 209
Neurofibromatosis, 211, 212
Neuro-metabolic disorders, 205
Nuclease, 42
  restriction, 60, 109, 110, 167, 169
Nucleic acid (*see* DNA and RNA), 78, 92, 162
Nucleotide, 33–35, 41, 42, 46–50, 158
Nucleus, 184

Open reading frame (orf), 163
Opportunistic infections, 92
Organ, 53, 82
  transplant, 53, 94
Organic acidemia, 207
Orton dyslexia, 200
Osteogenesis imperfecta, 208, 209
Osteoporosis, 141

Paget disease, 209
Parkinson disease, 210–12
Phage Group, 23, 26 30, 31, 36
Phages (*see* Bacteriophages)

Phenylketonuria (PKU), 15, 68, 69, 148
Phosphate, 158–60
Plant engineering, 143–46
Plasmid, 162–64, 169
Polyethylene glycol, 78
Polymerase chain reaction (PCR), 171–73
Polyposis, 195
Porphyria, 205
Prader-Willi syndrome, 197
Primer, 171–73
Progeria, 200
Promoter, 79, 80, 165
Protease, 185
Protein
  DNA-binding, 160
  harmful, 96, 97, 174
  modification, 133
  structure, 43–45, 48
  synthesis, 48, 49, 162, 165
  types, 53

RFLP, 60, 73
RNA
  antisense, 83, 174, 175
  messenger, 49, 79, 160, 164, 165, 174
  polymerase, 50, 164, 165
  structure, 48, 49, 164, 165
  transfer, 51, 162, 165
  viral, 78
Rabies, 146, 147
Recombination, 27, 28, 46, 61, 62, 66
Reflex sympathetic dystrophy syndrome, 212
Repressor, 165
Restriction endonuclease, 60, 109, 110, 167, 169
Restriction fragment, 60, 109, 111–13

Restriction fragment length polymor-
phism (RFLP), 60, 61, 109
Retinoblastoma, 139
Retrovirus, 78, 79 (*see also* HIV)
Rett syndrome, 211
Reverse transcription, 164
Ribose, 164
Ribosome, 49, 51, 164
Ribozyme, 83, 174, 175, 181
Rubenstein-Taybi syndrome, 200

Schroedinger, Erwin, 24, 30, 31, 33
Scleroderma, 216
Scoliosis, 208, 209
Sickle-cell disease, 15, 84, 148, 203
Sickle-cell trait, 17
Sjogren syndrome, 204
Southern blotting, 64
Spasmodic torticollis, 211
Sperm, 62, 83, 146, 179
Spina bifida, 212
Spinal muscular atrophy, 214
Sputum, 101
Stanley, Wendell, 26
Staphylococcus, 96–98, 103, 163
Stem cells, 83
Stent, Gunther, 30, 49
Sturge-Weber disease, 212
Sugar, 158, 161, 164
Superoxide dismutase, 132
Symonds, Neville, 38

Tay-Sachs disease, 14, 16, 85, 148, 150,
207
Thalassemia, 50
Thrombocytopenia abset radius syn-
drome, 203
Thymine, 35, 48

Tissue, 52, 53
incompatibility, 77
transplant, 77, 94
Tomato engineering, 143–46
Tourette syndrome, 53, 213
Toxic shock syndrome, 96–98
Transcription, 48–50, 162, 165
Translation, 49, 51, 165
Transposon, 98, 163, 164
Tuberculosis, 52, 92, 99, 100, 101, 103,
106
Tuberous sclerosis, 212, 213
Turner syndrome, 197

Ultrasound, 11
Umbilical cord, 83
Uracil, 48

VNTR, 109–18, 120, 124
Vaccinia virus, 146, 147
Virus, 44, 48, 78, 82, 90, 99, 158
bacterial (*see* Bacteriophage)
cowpox, 146, 147
plant, 20, 26, 145
RNA, 164
rabies, 146, 147
slow, 129
therapy, 174, 175

Watson, James, 1, 31, 33–35, 45
Wexler, Milton, 59
Wexler, Nancy, 59, 64–67, 73
Wilkins, Maurice, 33
Williams syndrome, 207
Wilson's disease, 206, 208

X-ray, 31, 33, 54, 101, 111, 112, 158
Xeroderma pigmentosum, 216